MARK HENRY SEBELL

WITH JEANNE YOCUM

A **Kaplan Professional** Company

This publication is designed to provide accurate and authoritative information in regard to the subject matter covered. It is sold with the understanding that the publisher is not engaged in rendering legal, accounting, or other professional service. If legal advice or other expert assistance is required, the services of a competent professional should be sought.

Publisher: Cynthia A. Zigmund
Acquisitions Editor: Jean Iversen Cook
Managing Editor: Jack Kiburz
Project Editor: Trey Thoelcke
Interior Design: Lucy Jenkins
Cover Design: Scott Rattray Design
Typesetting: the dotted i

Printed in the United States of America

01   02   03   10   9   8   7   6   5   4   3   2   1

Library of Congress Cataloging-in-Publication Data

Sebell, Mark Henry.
    Ban the humorous bazooka / Mark Henry Sebell, with Jeanne Yocum.
        p.   cm.
    Includes index.
    ISBN 0-7931-4108-7
    1. Technological innovations—Management.   I. Yocum, Jeanne.   II. Title.
HD45 .S396 2001
658.5'14—dc21

                                                                                00-011408

Dearborn Trade books are available at special quantity discounts to use as premiums and sales promotions, or for use in corporate training programs. For more information, please call the Special Sales Manager at 800-621-9621, ext. 4514, or write to Dearborn Financial Publishing, Inc., 155 North Wacker Drive, Chicago, IL 60606-1719.

Praise for *Ban the Humorous Bazooka*

"*Ban the Humorous Bazooka* focuses on how companies can enable all employees to participate in innovation. This truly is a practitioner's guide for leaders and managers determined to make a lasting contribution to the organization."

C.K. Prahalad, coauthor of *Competing for the Future*

"As a research analyst steeped in data about consumers and dreaming of creative applications to make the numbers come alive, I found [that] Mark Sebell's book *Ban the Humorous Bazooka* hit the sweet spot. 'Just the facts, ma'am' doesn't work for me as a personal credo or as a way to help my clients develop successful marketing strategies. Mark Sebell's book goes beyond the facts. His wonderful stories humanize the concepts and the tools necessary to merge innovative thinking and research. It's a winner."

Barbara R. Caplan, Partner, Yankelovich Partners

"An excellent guidebook for the innovator. Innovation is not only important in business but in other types of organizations, such as nonprofits, schools, and government agencies. Leaders of those organizations will learn useful lessons from this book."

Arthur Nelson, President, American Innovation Institute, Inc.

"This is as close to an innovation cookbook as I have seen."

Laurie LaMantia, coauthor of *Breakneck Teams for Breakneck Times* and adjunct professor, DePaul University

"I have witnessed first hand the creative breakthroughs that can be achieved by utilizing the principles and techniques described in Mark's new book. I now keep a toy bazooka prominently displayed in my office as a constant 'innovation reminder' that every idea, no matter how unconventional or implausible, has potential value and should always be encouraged."

Alex Perriello, President and CEO, Coldwell Banker Real Estate Corporation

"This is a practical manual on the how-to of innovation. From Miss Frizzle's advice to 'take chances . . . get messy . . . make mistakes,' right through to the criteria for a successful innovation launch, you'll enjoy these entertaining, experience-based concepts, including: how to put together champions for innovation, to warding the dangers of speed bumps, to interactive consumer dialogues."

Joan Bonnette, VP and Marketing Director, Retail Banking, Citicorp

"Too many business books talk about what companies should do in theory. Mark Sebell has a genius for transforming theory into useful ideas. He provides lots of practical ideas, entertaining stories, and clear examples that you can use immediately to actually improve the innovation process in your organization."

Dr. Deanna Berg, President, Innovation Strategies International

"Organizations that have recognized the growing war for talent typically confront the battle on two fronts: They either seek to improve their acquisition practices to make their companies more attractive to top candidates and/or adopt developmental programs to optimize their internal talent. Mark Sebell's insights and processes for innovation can help you do both. Innovative companies are clearly more attractive places to recruit talent to, but perhaps more importantly, the people that currently surround you, regardless of [their] background, skill, position, [and] experience, are an extraordinary and likely untapped source of creativity that could lead to business-building innovation. *Ban the Humorous Bazooka* can show you how."

Jose Aragon, Vice President Marketing, Spherion Corporation

"I have worked with Mark Sebell to deploy these concepts and processes at three market-leading, high-tech firms over the last ten years. Perhaps one of the most important learnings for me is how the approach moves the entire team toward innovation, a prerequisite for success. Everyone has a role and an important contribution to make. So it is with innovation— not everyone is the creative genius, but the larger part of your team must be involved, each playing their specific role. And that is where Mark's approach is valuable beyond all others—all key stakeholders are *enabled* to move forward together."

Joe Pajer, Executive Vice President and General Manager,
Routing & Switching Products, Marconi

Dedicated to:

Amy, Jason, and Rachel Sebell
Who made the big sacrifices
In support of my passion
For this line of work

Abe Horowitz, Harry Sebell, and
Norman Sebell,
My entrepreneurial inspirations

The memory of
Kathleen Campbell Yocum

The strategic core purpose of this book is:

To improve the innovation success rate of business

Contents

Phase Three: Invention

Phase Four: Greenhouse

Phase Five: Implementation and Launch

Acknowledgments

From Mark Henry Sebell

An entire lifetime of experiences is reflected in the stories, conclusions, and recommendations you will find in this book. Many people have contributed to these learnings. In particular, I want to thank all of those people mentioned in these stories for allowing me to use their names, companies, and projects to illustrate some of the more significant concepts.

In the course of writing this book, I happened upon two works that helped me clarify some of the principles underlying the theories put forth here. These books are: *Built to Last: Successful Habits of Visionary Companies* by James C. Collins and Jerry I. Porras; and *Management of the Absurd: Paradoxes in Leadership* by Richard Farson. I encourage you to read both of these books; they each contain numerous ideas that will be helpful before you begin your journey on the innovation highway.

As stated in the Dedication, I want to thank my wife Amy, son Jason, and daughter Rachel for enduring years of my absence while I've been on the road pursuing my passion for this line of work. They have given greatly of themselves.

Also, to my colleagues at my company, Creative Realities, whose contributions have been intricately woven throughout this compendium of stories and skills, I must say, "Thanks to one and all." I especially want to applaud one of my partners, Cris Goldsmith, who has weathered many an adventure with me and been the perfect counterbalance to my rather voluble personality.

My other business partner, Brian Walker, read every word of at least two drafts of this book, offering many constructive "builds" that have made their way into the final edit.

An enormous debt of gratitude is also owed to my cowriter, Jeanne Yocum. Having written articles and other materials together for many years, she encouraged me to consider coauthoring a book when it was the last thing on my mind. I will never forget signing our agent's contract in the grandstand at Fenway Park on a bright, late October afternoon, just before the 1999 baseball playoff game between the (dreaded) New York Yankees and our (beloved) Boston Red Sox (Pedro Martinez beat Roger Clemens). Ever since, her gentle prodding and profound energy have kept me going through our "Dark Night of the Innovative Writer." We never fought, though we often disagreed. And, along the way, she has become a true friend. If this book is at all readable, informative, and enjoyable, it is in large part due to her. She is the best writing partner on a very difficult topic that anyone could wish for.

Finally, I want to thank my friend and fellow Colgate Thirteener (college singing group) Scott Christensen, who thought he was coming to Nantucket for a mini-vacation and ended up spending three days proofreading and strengthening every chapter of this book.

From Jeanne Yocum

In 1994, two people, Katherine Catlin and Kathy Leydon-Conway, recommended that I attend the Creative Problem Solving Institute (CPSI), held each June in Buffalo, New York, by the Creative Education Foundation. I value the opinion of both women, so I went to CPSI, had a transforming experience, and ended up going back for the following two years. Thus it is that I am grateful to each of them for starting me down a synchronistic path that prepared my mind and spirit to take on the challenge of coauthoring this book. Similarly, heartfelt thanks go to that best of friends, Joan Schneider, for introducing me to the people at Creative Realities.

Finally, to my fellow traveler on this often mysterious, sometimes frustrating, mostly thrilling ride of coauthorship, all I can say is thanks, Mark, for making the leap of faith. One lifelong dream has now been fulfilled (to have my name on the cover of a book). Now, if only the Red Sox would win the World Series!

Joint Thanks

We both extend to our agent, Nicholas Smith of Altair Literary Agency, an enormous thank you for seeing the potential in our early proposal, for patiently answering our neophyte questions, and for being a master negotiator on our behalf. Also, our gratitude goes to the many people at Dearborn Trade who contributed their energies to this book, especially our editor, Jean Iversen Cook. Finally, we thank the *Design Management Journal* for allowing us to adapt materials from an article we wrote that originally appeared in their Fall, 1997 issue.*

Note

Design Management Journal, Fall 1997, a publication of the Design Management Institute, 29 Temple Place, Boston, MA 02111. Ph: 617-338-6380/fax: 617-338-6570/<www.dmi.org>

Foreword

by C. K. Prahalad

Two issues are becoming clear in the emerging business landscape, and old ways of doing business will not be sufficient to address them. First, relationships with suppliers and customers are changing radically. The combined forces of deregulation, industry convergence, and the emerging markets are creating a new competitive reality. Second (and perhaps more importantly), the Internet is forcing a revolution in the way we think about products and services and the relative roles of the firm and the consumer. Strategy in this context is not about being efficient in a given paradigm but in inventing a new one.

Until now, strategy has been primarily thought of as a process of positioning a company in a defined industry space. The focus has been on increasing efficiency and tweaking products based on fairly reliable information about competitors' resources and capabilities. This game has been well understood by all the players.

Increasingly, however, strategy is about inventing a new game. While the old strategic levers—costs, cycle time, and quality—are still critical to a company's success, the degree of innovation is increasingly what distinguishes companies with superior performance. Thus far, start-ups have enjoyed a reputation as centers of innovation, and little attention has been paid to reinventing traditional and well-established firms. In this book, Mark Sebell's critical contribution is the argument that innovation is a cultural component that must be embedded in organizations of all sizes. In addition, he addresses many of the basics on creating that culture in simple and easy-to-understand language.

The new economy is about innovation and the search for talent worldwide. Many companies are competing for talent; *Ban the Humorous Bazooka* focuses on how companies can enable all employees to participate in innovation. Innovation is not about "skunk works" or a departmental function; it is concerned with the total organization. Further, innovation must change from a sporadic and reactive effort to a cultural phenomenon. While empowerment has become a buzzword in many circles, many of the significant questions that naturally follow (How fast? How can efforts be aligned? How should these efforts be rewarded?) have not been sufficiently addressed. This book presents an extremely well-informed view of how to make innovation a foundational element of the culture of any organization.

In addition, *Ban the Humorous Bazooka* demystifies many of the intermediate steps along the path of creating the culture of innovation. Managers tend to either make it a formalized process that sometimes lacks intellectual rigor (easy to replicate across organizations, even if the idea has no merit) or they allow it to be a mystery (it may happen spontaneously and unpredictably, if the employee can get someone's attention). Neither of these extremes creates winning companies. Although the anecdotes in this book are easy to read (and recognize), don't be fooled by the easy read. Embedded throughout is a deep understanding of the theory of an innovative culture in a large and established organization.

Mark Sebell makes many useful distinctions. For example, creativity is not innovation. Innovation is, to borrow from Karl Weick, a process of disciplined imagination. Further, there is not one standard process of innovation. Many of the impediments to innovation are likely to be unintentional—like humorous bazookas. Often, even those who are putting speed bumps and roadblocks in the path of change and innovation do not realize that they are doing so. The innovative culture creates a new kind of company; one focused on harnessing chaos. Such a culture abhors hierarchy. The basis for influence and power is knowledge and contribution. It is also likely to represent ambiguous and ever-changing internal environments. Those who abhor ambiguity will find these organizations to be difficult.

This book identifies the basic intellectual and organizational building blocks of innovation. The focus is on: "How do I do it?" As a result, it has one of the most interesting characteristics for a book on innovation; it is as much about individuals and their values as it is about the genetic code of the total enterprise. It focuses on the individual, the small group, and the total organization.

If you are a manager struggling with the reality of the New Economy and feeling the pressure to innovate, you will find this book valuable. Mark Sebell's many examples illustrate his concepts clearly. This truly is a practitioner's guide for leaders and managers determined to make a lasting contribution to the organization.

Competing for the future is all about stretching our imaginations and shaping consumer expectations. If you are concerned about effective ways to change the world (starting in your own organizations), you would do well to get started here.

C.K. Prahalad
Chairman, Praja, Inc.
Harvey C. Fruehauf Professor of Corporate Strategy,
The University of Michigan Business School

Introduction

hu-mor-ous ba-zoo-ka (hew'mer-us be-zoo'ka), n. 1. a funny, witty comment that, intentionally or unintentionally, shoots down another person's idea. 2. innovation killer.

If you've elected to read this book, I assume you have already accepted the proposition that continuous innovation is of paramount importance in today's constantly evolving marketplace.

If, on the other hand, you don't agree that innovation is the very lifeblood of successful business, then you might want to return this book to the store where you purchased it. Convincing readers of the link between innovation and a healthy bottom line is not what I intend to accomplish here. However, if you are certain of the necessity of continually creating and implementing innovative new products, new services, and new ways of doing business, but are puzzled about how to achieve that goal, I encourage you to read on.

A strong belief in the indispensability of innovation (beyond mere creativity) is a prerequisite for successfully navigating the innovation highway. Learning how to master innovation is not easy. It is definitely not child's play although, delightfully, parts of it are as enjoyable as child's play. But mostly it is hard and challenging work. If you believe that constant innovation is the only way your organization can flourish over time, you may survive what people in my field call the "Dark Night of the Innovator."

This is when human behavior creates obstacles that inevitably slow down—or even kill—successful innovation.

This book details much of what you must know to take a successful journey on the innovation highway. In it you will learn about these seven concepts:

1. *The difference between creativity and innovation.* The exciting, new idea is a rare and beautiful thing. But, too often its creation is seen as the beginning and the end of an innovation initiative. This is a mistake made in far too many organizational environments, and as a result most of those great ideas never see light of day. This book begins with an explanation of the difference between the intricately linked but very different concepts of creativity and innovation.

2. *The levels of innovation.* There are three levels of innovation that you can pursue. Each embodies a different element of risk, often unforeseen. You will learn about each, how to define them and how to develop a portfolio mentality in laying out your roadmap to innovation.

3. *The phases of the innovation process and the role each one plays in achieving successful outcomes.* Innovation requires the acceptance of many paradoxes. One of them involves the premise of *flexible process,* the management of innovation in an organized, yet adaptable way. This book presents a five-phased flow for achieving innovation. Each phase is important and each provides a foundation for the one that follows it. While it is tempting to think you can speed up innovation by skipping ahead in the process, you will miss something vital if you do so. By understanding the possibilities that can be identified by thoroughly exploiting each phase, you will be more likely to achieve breakthrough, even transformational, innovation.

4. *The critical success factors that must be present in each phase of innovation.* In addition to my 30 years of working in the innovation field, my colleagues and I have identified certain conditions that companies that master innovation enjoy in common. Conversely, these are conditions that are usually lacking in organizations that fall short when it comes to being innovative.

5. *The common mistakes that cause innovation efforts to fail.* There are speed bumps (individual behaviors) and roadblocks (group behaviors and cultural norms) that impede your progress as you work to bring about change.

6. *Tools and techniques that will help you overcome innovation speed bumps and roadblocks.* You will learn new skills and behaviors that will make

you and your organization better equipped to win the innovation battle. These include ways in which you can help alter individual behaviors as well as significant cultural changes that can remove huge organizational impediments to innovation.

7. *Innovation Fuel.* Each chapter of this book concludes with "Innovation Fuel," a summation of cornerstone concepts from the chapter that will give you and your team the power needed for a successful trip on the innovation highway. As you find yourself bumping up against barriers or suffering in the Dark Night of the Innovator, you can return to these short sections for quick insights into what might be going wrong and for ideas on how to overcome problems.

As you seek to create a culture that nurtures ongoing innovation, perhaps the most important caveat to bear in mind is that innovation is messy. You have got to accept that fact and be prepared for what it means in terms of frustration, disappointments, and even the occasional broken heart, which happens when you pour your soul into something that may ultimately fail.

Many people have tried to eliminate the messiness and risk inherent in innovation by developing step-by-step, so-called innovation processes that can be shoehorned into any business. The reality is that no one process will work repeatedly even in a single organization, let alone multiple organizations.

At its core, innovation is about people. My goal with this book is to help business people understand and appreciate how to work effectively together in ways that support breakthrough innovation. My colleague, Frank Hines, loves to quote Miss Frizzle, who advises the children who watch TV's *The Magic School Bus* to "take chances . . . get messy . . . make mistakes." I hope you will be stimulated by this book to do just that.

Establishing the Right Foundation

CHAPTER ONE

Nine Myths of Innovation

This is a book about innovation in business. It is *not* a book about creativity in business. Creativity is the act of generating a new idea. Innovation involves making that new idea *real*. One of my partners, Cris Goldsmith, has defined it most succinctly. He says, "Innovation is creating and successfully implementing new ways of doing things."

Unfortunately, innovation happens all too rarely in most organizations. Understanding this fact is your initial step toward a new way of thinking. It will improve your ability to successfully travel the hazard-strewn road to innovation.

Creativity involves the much-clichéd act of thinking outside the box. And believe it or not, creativity happens every day in business. In fact, new ideas are a dime a dozen in most companies. Even the most stolid, tradition-bound organizations have individuals who are capable of looking at situations in a new way and saying, "Hey, here's a way we could do this faster/cheaper/easier/better."

What happens next with that blossoming idea determines whether or not a company is innovative. Most often, all sorts of obstacles arise whenever a company starts down that twisting path to turn a new idea into a real new product, a service enhancement, a technology improvement, or any consumer-focused innovation. The next thing you know, that fresh new idea is road kill.

To be innovative, a company must be able to manage and nurture the development of any new idea through the evolutionary maze to market launch. This is completely different from being creative, although the two are obviously related. To paraphrase a statement I once heard, you can have creativity without innovation, but you can't have innovation without creativity.

Creativity is all around us. Innovation is, too, but innovation is infinitely harder to achieve, particularly in a business environment.

Creativity tends to spring from individuals who have an instinctive ability to see new possibilities. In contrast, business innovation is accomplished by groups. It requires visionary, cross-functional teamwork—teamwork that is focused, in part, on stretching and even breaking the rules. And, while business leaders actively espouse a passionate conviction that innovation is the cornerstone of their economic growth, most have failed to put in place the culture and processes that will produce innovation. They have misunderstood the real meaning of cross-functional teamwork and have not fostered working environments that allow rules to be bent, let alone broken.

Is This Your Company?

Think back to the last brainstorming session in which you participated, where the goal was to come up with innovative ideas for your business. How many creative ideas were put forth for the group's consideration? How many of those really new ones survived the barrage of negativity and doubt that usually greets new concepts? And, of the ideas that did survive, how many have been implemented or are still moving in that direction? Very likely, few made it into development and fewer still—if any—actually are on their way to market. You've got the creativity part down; you just haven't learned how to be innovative!

If your company is typical, I'll wager that plenty of good ideas surfaced during the brainstorming but few, if any, of the truly breakthrough ones made it out of the room alive. Most of the truly new ideas were probably shot down with a barrage of humorous bazookas—the act of shooting down another's idea with a witty barb.

This tendency to lob verbal grenades at new ideas has been and still remains so pervasive that I coined the term *The Bazooka Syndrome* in 1982, when I first began my career as a creative problem-solving facilitator. Every time I have described this behavior to a new group of people, it has hit a responsive chord. Everyone instantly identifies with The Bazooka Syndrome because we all have been hit by these verbal missiles. And most people will also admit, with shamed faces, that they have been guilty of using bazookas on the ideas of others (colleagues, spouses, kids, family, and friends).

The Bazooka Syndrome captures what we unintentionally, but instinctively, do to new ideas. We make fun of them. We point out every single problem. We end up annihilating them, all in the spirit of constructive flaw-finding and, allegedly, idea improvement.

For creative people who are good at generating fresh ideas, being hit by a bazooka blast is enormously discouraging. Frustration abounds in organizations that are skilled at dreaming up new ideas yet ineffective at protecting them from the bazooka wielders that exist everywhere.

It's very discouraging to watch competitors successfully launch innovations based on ideas you tossed around but failed to pursue because you were gunned down by a bazooka. Are the phrases "Gee, we thought of that months (or years!) ago" and "We tried that but couldn't make it work" commonly heard within the walls of your organization? If so, your company is undoubtedly populated by bazooka experts and, as a result, is short on innovation.

Why Is Innovation So Hard?

Here's a sad fact: At least 25 percent of the consulting work my company performs involves helping teams implement and launch innovations they've been contemplating for a long time. The problem is not in generating ideas. The problem is a lack of ability to implement ideas—not knowing how to overcome flaws in the ideas and not knowing how to navigate those ideas through an organizational structure that doesn't welcome change.

Why is it so challenging for people and entire organizations to embrace innovative ideas and then do the hard work that is required to turn them into reality? Perhaps this reluctance to innovate is an unintended legacy we inherit from our school years.

George Bernard Shaw said, "All great truths begin as blasphemies." Having sat through history and science classes in junior and senior high school, we can all cite a lengthy list of people who faced tremendous struggles—even risked death—because they pursued truly revolutionary, new, transformational, "blasphemous" ideas, like:

- Abraham, who believed in one god in a time of polytheism,
- Copernicus, who waited 13 years before feeling safe enough to publish his theory proving that the sun, not the earth, is the center of our solar system,
- Charles Darwin, who is still derided by some nearly a century and a half after he put forth the theory of human evolution, and
- Freud, who endured vilification when he began to probe the psychology of the human mind.

After studying about person after person who faced dire consequences for supporting ideas labeled extreme (blasphemous), perhaps the strongest lesson we (subconsciously) learn is that being innovative is not such a great idea. By the end of high school, most of us have internalized the belief that dedicating our lives to pursuing new ideas is mighty risky business.

One outcome of this collective reluctance to crawl out on the limb in support of anything new is that, in the business world, being innovative is one of the toughest challenges we face. Safely birthing a new idea—bringing it from the state where it is just a new, imperfect notion to the point where it is a legitimate new product, a valuable service improvement, or just a better way of doing business—is indeed a struggle.

In fact, in most organizations, innovation initiatives are usually perceived as wasted time and money down the drain. Too often the only results are missed opportunities that result in frustrated staff, disappointed senior management, and no bottom-line improvement. These judgments miss the point because one of the essential differences between breakthrough innovators and everybody else is that, for them, there is no failure—only learning.

I remember, as a young boy, watching Spencer Tracy playing Thomas Edison in a movie. The scene was in Edison's Menlo Park laboratory where he and his fellow workers were experiencing the frustration of another failed attempt to develop the right filament for an incandescent light bulb.

In total disgust, one of the lab people muttered that "it's just another one that didn't work." Wise old Tom Edison looked bewildered because his colleague didn't understand that "now there are a thousand that we know won't work."

Innovation Is Scary Stuff!

Another reason innovation is difficult to implement is because it requires change and plenty of it. Attitudes and systems must change. People must learn and internalize creative thinking and problem-solving skills. Cross-functional departments must team to work together differently. These new concepts must be adopted by the organization rather than ignored, and all of this must happen while everything that supports the old way of doing things fights the change.

Innovation frightens most people. As 19th-century English economist Walter Bagehot observed, "One of the greatest pains for human beings is the pain of a new idea." No wonder people raise objections, dig in their heels, throw up their hands, and say, "It can't be done."

The innovator lives and relives the Dark Night of the Innovator (a phrase I will use often in this book), which is that place and time in the innovation process where implementation seems so remote that you doubt its possibility. This is where you have to constantly buck the status quo, leap over hurdle after hurdle, and, perhaps most challenging, struggle to keep faith with your idea and your vision of what might be. This is hard work!

Even in situations in which everyone honestly gives their best effort, the results often fall short of the objectives. The innovation team ends up tinkering with and modifying what exists instead of seriously considering what might be. They deliver me-too ideas instead of the desired breakthrough innovation.

Damaging Myths about Creativity and Innovation

I have found that even people who can distinguish between creativity and innovation and who aren't intimidated by how difficult innovation can be still cling to common myths about how creativity and innovation work. These misunderstandings crop up at all stages of innovation, causing confusion and impeding progress. It is important that you free yourself from these misconceptions immediately so that they don't color your viewpoint later, when I begin to discuss how innovation actually works.

Myth 1: You Can Purchase an Innovation

New products and new services have always been two of the most visible ways for companies to innovate. But, they are not the only means by which your company can compete. Important innovations can succeed in many other forums, including some that consumers will experience but may never actually see. Potential areas for innovation include:

- Manufacturing cost reductions
- Warehousing and distribution efficiencies
- Customer service improvements
- Creative marketing practices and promotions
- New forms of packaging

Last year my partner, Brian Walker, assisted a large computer manufacturer in identifying over $100 million in manufacturing cost reductions. These improvements were realized primarily inside the computer, where the consumer will never look. Yet, these innovations in production efficiencies were very meaningful and will help this high-tech innovator compete more efficiently in that half-life arena.

Some innovations today, particularly on the Internet, are being provided free of charge. The fact that they are being given away—as part of a marketing strategy designed to achieve a goal such as capturing market share or attracting eyeballs to drive ad revenues—is as much an innovation as the primary product or service that is being offered.

In fact, as the lead-time needed to duplicate the innovations of others continues to shorten, new products will not be the competitive preemptors that they once were. And, as the world demands more standardization in its technological innovations, the value of technology as a long-term differentiator will diminish.

All of this bodes well for the innovator because there are so many ways to be competitive. In the years to come, new products and new services will only be some of the many ways in which companies will be able to differentiate themselves and compete effectively. Companies that master innovation in all aspects of their operations will be the leaders in such a world.

Myth 2: All We Need Are Some Good, New Ideas

A client once sent me an anonymous saying that sums up the difference between creativity and innovation: "A great idea is still just an idea until you make it real." Or as author Michael Schrage puts it in his book *Serious Play: How the World's Best Companies Simulate to Innovate*: "The real test of individuals and enterprises is not how much they know or what they know; it's what they do with their knowledge. Making a decision is not the same thing as implementing it."[1]

These quotes refer to the fact that, in business, creative ideas and knowledge are meaningless without implementation. An unimplemented idea becomes vapor. Too often, creativity is seen as the beginning and end of innovation, a misconception that impedes innovation from the very outset.

If you study innovative market leaders, you will find that many of them are no more creative than their competitors. What they grasp, that poor innovators don't, is that coming up with good new ideas still leaves you miles away from achieving innovation. Such leaders have mastered the critical part of innovation—the skills needed to steer fragile ideas over the barriers that block their implementation.

Myth 3: Once We Shout "Eureka!" We'll Be Done

Here's another difference between creativity and innovation that you need to understand if you're going to survive the Dark Night of the Innovator: The forces that drive creative people and those that drive innovators are different.

The passion for creators is the desire for that "Eureka!" moment—what you might call the mental orgasm that comes when you hit upon what you know is a fabulous idea. If your enjoyment is derived largely from blinding flashes of insight—the light bulb suddenly going on inside your head—you may be truly creative, yet have little patience for the long, arduous road to innovation.

Think of all the idea people you've known in your life—individuals who seem to have a new notion every day but, too often, never manage to do anything with their brainstorms. These people are like idea hamsters. They always seem to have their idea generators running. Such creators love the buzz that comes from generating ideas but usually aren't stimulated or interested in the messy, hard work of implementation. They tend to shout "Eureka!" only to jump ship when it becomes clear that making that same idea real is going to be harder than they first envisioned.

If, in contrast to these "Eureka!" junkies, your fulfillment comes from nurturing novel, fresh ideas through to reality, then you may be a born innovator. You're not driven by pride of authorship and you're not intimidated by the notion of needing to motivate a cross-functional team to overcome the speed bumps and roadblocks that impede innovation. In fact, you downright enjoy the challenge of traveling the winding road to successful implementation.

Thomas Edison also said, "Genius is 1 percent inspiration and 99 percent perspiration."[2] To be a true innovator, you must be prepared for the 99 percent of innovation that is perspiration.

Yes, some individuals embody both of these driving forces but they are extremely rare. The ideal innovation team, of course, meshes idea creators and doers—people who excel at implementation. Getting these two types of people to understand and appreciate each other's strengths is part of the challenge of building a successful innovation team, a topic I'll address in far greater depth in Chapter 5.

Myth 4: The Right Idea Will Come Out of Nowhere

The "Eureka!" factor itself is a myth. It is rooted in the story of Archimedes, who history claims climbed into a bathtub and shouted "Eureka!" as he suddenly realized that the physical principle of buoyancy and displacement was the way to determine the purity of the gold in King Herion's crown.

The usual interpretation of this story is that the solution came to Archimedes in a blinding flash out of nowhere. That's misleading. What Archimedes had was, in the words of educator John Saxon, "a well-prepared mind" (a topic I will explore in great detail in Chapter 10). Archimedes' wonderfully creative brain was able to solve a problem when it met with a proper stimulus—the bath water spilling over the edge of the tub as his weight displaced it.

Archimedes had been cogitating on this problem for some time, turning it over in his mind and studying it from all angles. Finally, as he stepped into his bathtub, in a wonderfully paradoxical state known as relaxed concentration, all the pieces of the solution came together. But, without all that earlier preparatory work resident in his well-prepared mind, he might never have had his "Eureka!" moment (which professionals in my business call making a mental connection).

People and teams that are skilled at making creative ideas a reality know not to expect the instant answers that are implied by the "Eureka!" myth—the notion that out of nowhere will come the right answer. Rather, they appreciate that if they keep working on a problem a solution will eventually come to them. Sometimes it will be piece by piece, and other times it may, in fact, appear quite suddenly. But, whether the solution appears quickly or in piecemeal fashion, it will only come because they have been doing the hard work that, miraculously, seems to gel into unexpected insight.

Myth 5: I'll Recognize the Breakthrough Idea When I First See It

The notion that any really breakthrough idea will quickly be recognized as such, when initially conceived, is false. It is, in fact, extremely rare that a breakthrough new idea is recognized for its brilliance when first uttered. This is because most people evaluate ideas at a fixed point in time, usually when we first hear them. It is only with the benefit of hindsight that we come to realize that an idea that was labeled stupid at first blush was, in fact, brilliant.

Delve into the story of how any number of fantastic new ideas made it to market, and you'll be hard pressed to find many cases in which the breakthrough ones were met with shouts of "Hallelujah" when first put forth. Instead, in most workplaces, breakthrough ideas are routinely greeted with bazooka blasts. It's only when someone else implements them that they look good.

Why do we more often than not reject sensational beginning ideas—concepts in their earliest stage of life—that turn out to be no-brainers when they have been fully developed? The answer lies in Albert Einstein's statement that "if at first a new idea doesn't seem totally absurd, there's no hope for it." And, as Thomas Carlyle noted, "Every new opinion, at its starting, is precisely in a minority of one."

How many of us are willing to support an absurd idea? How many trust our colleagues enough to join their minority of one when an idea they put forth is truly breakthrough, and thus, probably appears nonsensical? This is why brilliant ideas rarely are labeled as such when first introduced and why we may run the other way when faced with the beginning of a truly breakthrough concept.

This mistaken belief that you will instantly recognize a brilliant idea when you hear one is extremely damaging to an innovation effort because new ideas almost always are flawed in some way when they first appear.

Given the risk-averse mind-set that thrives throughout the average corporate culture, most people work in an atmosphere where volumes of beginning ideas, with brilliant potential, are ignored because their inherent value is not immediately evident. The commitment to Edison's 99 percent perspiration that is needed to make these ideas a reality simply doesn't exist.

Myth 6: To Be Innovative, We Need a Clearly Defined, Repeatable Process

Let's get this straight from the start: There is no single road map for innovation. Hundreds of millions of dollars have been spent by corporations in fruitless efforts to map what poet William Blake called "the crooked road . . . of genius." This search for an ordered, logical set of steps and procedures that will lead anyone and everyone to innovation overlooks the inherent messiness of innovation. Innovation efforts do, however, benefit from a flexible process approach, as you will learn in this book.

What you must do during the Dark Night of the Innovator, when things get seriously messy, is motivate an entire organization to play with a new idea. This is daunting because, until the idea begins to take on a life force of its own, you don't get rewarded for working on it. It's just one more thankless innovation activity that is usually relegated to the hours after your regular workday is done.

A Taste of a True Innovator

Even 3M, with its famous bootlegging policy that strongly encourages scientists[3] to spend up to 15 percent of their time in the unstructured pursuit of projects they personally find intriguing, doesn't have a rigid roadmap to guide innovation. The story of how Art Fry, the inventor of Post-It Notes, had to fight to get his idea off the ground at 3M demonstrates this. While most people have heard different versions of how he created the idea, few have heard what had to transpire for it to become a successful innovation.

In 1973, Fry attended a seminar at which 3M research scientist Dr. Spence Silver discussed how in 1968, while studying a totally different issue, he had developed an odd adhesive that formed into tiny spheres that were very sticky individually yet did not stick strongly when coated onto tape backing.

For five years, Dr. Silver had been discussing this strange adhesive with others at 3M, trying to figure out what to do with it. When Fry saw Dr. Silver's presentation, he saw this nonpermanent adhesive as the answer to a problem that had long plagued him—how to insert location tabs in his choir hymnal without damaging the book. Initially he thought of this new product as a great bookmark but from there he quickly formulated the concept for Post-It Notes.

However, when Fry presented the idea to his management, a number of objections arose as others didn't appreciate the potential and rejected it— an excellent example of a great idea whose value went unrecognized, at first, by others. Engineering and production said the product would be too difficult and expensive to manufacture. Marketing believed that nobody would pay money for something that appeared to do the same thing as cost-free scrap paper.

Rather than let the concept die, however, Fry used his bootlegging time and devoted 18 months to solving the technical problems and then persevered until he was able to convince department after department that the idea would work. Still, initial test marketing didn't go well until an analysis showed that some dealers were enjoying extraordinary sales. It was soon determined that in stores where free samples were given away the product took off like a rocket. Once people experienced Post-It Notes, they purchased the product. It is now one of the top-five-selling office products in the United States.

The moral of this story is that even at a company as philosophically committed to innovation as 3M, worthy ideas like Fry's have to fight for survival. Even 3M has no magical implementation hopper into which innovators can deposit ideas that automatically come out the other end ready for market. No such protocol exists anywhere, at least to my knowledge.

What 3M does have in place is a core value and a human resource philosophy that allows people a certain allocation of time to devote to innovation. I think the fact that there is no real structure to their 15 percent policy is exactly why it works.

It's a true people-oriented process that is akin to the January Study Program that existed when I was an undergraduate at Colgate University in Hamilton, New York. Every student spent the entire month of January working (pass-fail) on virtually anything he or she wanted to study. In that month, half of my classmates produced little of value (which is also probably true at 3M). Others may have abused it somewhat or even entirely. But, it's what happened with the other half that leads to breakthroughs whether in academia or in business.

Myth 7: Innovation Has to Be a Home Run

Here's where the Japanese really understand one of the vital principles of business innovation. For decades, they were content to introduce small and medium-sized innovations, one after the other, across a range of industries, particularly in electronics and automotive. In baseball terminology, they singled and doubled competitors to death.

In the rest of the world, however, businesses have too often resisted change and newness of any kind until they realized they were lagging way behind the innovation curve. They then decided they needed to hit a home run to catch up. In this state of panic, if ideas that are put forth aren't gigantic, breakthrough concepts, they get rejected. Because big, new ideas tend to appear blasphemous and/or seriously flawed, the panic can build to hysteria (usually with lots of finger pointing and "blamestorming").

Let's explore that home-run mentality for a moment to understand why it is too risky to be your company's innovation strategy. How many Babe Ruths and Hank Aarons appear in one lifetime? Not many; the last century produced only two, although Sammy Sosa and Mark McGwire aren't through yet. More important, please note that, while sluggers Sosa and McGwire each had two consecutive, 60-homer seasons in 1998 and 1999, neither one's team made it close to the World Series in those years. The lesson here may be that, while home runs make stars of individuals, the goal of *only* home runs usually ends in less-than-stellar results for a team.

Myth 8: Innovation Can Be Accomplished in One Meeting

I am constantly astonished when clients expect to define a comprehensive future for their companies, divisions, or brands in a morning or a day. This is a completely impossible objective that is steeped in the confusion over creativity versus innovation. Creativity might be achieved in one meeting, but innovation requires an unpredictable number of interactions that bring together groups whose composition changes, based on where you are in the flow of innovation.

If you're acting on the no-brainer ideas that arise in every brainstorming session, you may agree to implement them immediately and identify the appropriate people to just go and do them. But if you are striving for something that truly will reinvent your company, don't make the mistake of believing that such transformation will magically start to happen the minute you leave any brainstorming session. Instead, prepare yourself for a trip through the Dark Night of the Innovator.

Myth 9: We Just Implemented a Great New Idea; We Can Rest Now

It is critical to understand that innovation is continuous. You cannot say, "Okay, we've innovated; now we can sit back and watch our profits grow." Remember the lesson that was learned from Tom Peters' and Robert Waterman's *In Search of Excellence: Lessons from America's Best-Run Companies.* Many of the companies labeled as excellent in that book became so complacent that, within a very short time, they found themselves in trouble. If you take a let's-rest-on-our-laurels approach, someone will come along and start grabbing your market share before you have the time to restart your innovation engine.

Today's marketplace is dynamic and constantly changing. To respond—indeed, to stay in business—you must foster a culture that understands the need for continual change. Although many know this intuitively, too few organizations respond in an effective or timely manner. Organizations that master the art of continuous innovation are the ones that win the competitive war.

Kill the Myths

The nine myths discussed in this chapter cause many innovation efforts to die on the vine. Understanding that they are, in fact, myths gives you a more realistic appreciation for all the arduous work that lies ahead as you pursue innovation. Put these myths to rest by developing a better understanding of the nature of innovation—including how it differs from creativity.

[i n n o v a t i o n   f u e l]

- Appreciate the fact that having a handful of good ideas doesn't mean you're anywhere close to achieving innovation. In fact, you will have only just begun.

- Understand that you need an innovation team that mixes "Eureka!"-driven, creative people with implementers who are willing to make the long, difficult slog through the Dark Night of the Innovator.

- Avoid sitting around waiting for solutions to hit you out of nowhere. Instead, learn how to be constantly vigilant about exploring and probing for sources of new idea stimulation.

- Do not assume that you will recognize a breakthrough idea the instant it's proposed. Learn how to explore the potential value of *all* beginning ideas and be willing to give scary ideas time to grow.

- Do not waste time and money on seeking the perfect innovation process because one, single, repeatable road map to innovation doesn't exist. Instead, adapt a flexible process that can be altered as your innovation objectives and your competitive situation change.

- Increase your batting average by broadening your innovation strategy beyond the we-need-a-home-run-every-time mentality.

- Accept that you can't reinvent your company (or division or department) in a day. It is folly to expect to achieve anything this critical in such a short time frame.

- Do not innovate in fits and starts; build a culture that encourages continuous innovation.

- Recognize that your search for innovation should not be confined to something that will sit on a store shelf. Opportunities for innovation exist in every part of your business, not just in the product or service development department.

Notes

1. Michael Schrage, "Play Power," *Business 2.0*, November 1999, pgs. 261–274 (excerpted from *Serious Play: How the World's Best Companies Simulate to Innovation*, Boston, Harvard Business School Press, 1999).
2. M.A. Rosanoff, "Edison in His Laboratory" (4), *Harper's*, September 1932, quoted by Leonard Roy Frank, *Influencing Minds*, Portland, OR, Feral House, 1995, pgs. 181–182.
3. 3M also allows nontechnical employees to devote up to 15 percent of their time to working on an intriguing idea, with supervisor approval.

CHAPTER TWO

What You Must Accept about Innovation

Understanding the unique, often contradictory (paradoxical) challenges of the innovation process is critical to the success of any endeavor designed to produce newness. One immediate truth, which can be extremely difficult for many business managers to accept, is that the thinking skills and decision-making structures that are effective for pursuing innovation are different from those needed to run an existing business.

Pursuing innovation—especially breakthrough innovation—with the same mind-set that you use to operate your current business is almost guaranteed to end in frustration and failure. This is why adopting the mind-set of an intrapreneur (someone who thrives on being entrepreneurial within a large, established organization) is so important. The intrapreneur instinctively develops unconventional approaches for bringing innovation to life in an existing organization.

To help you begin to understand why innovation is so difficult to achieve, here are six critical lessons about the nature of innovation that you must learn before accelerating onto the innovation highway.

Six Critical Lessons about Innovation

1. Innovation Is a Trip to a Vague and Uncertain Future

Innovation is a journey into the unknown. Certainly, you must start with a clear objective in mind but you can never say at the outset exactly what your final destination will be. And yes, you try to plan the best possible route before setting out. But once you're underway, interesting side-trip opportunities will present themselves and you may decide to take some detours.

Innovation is not a linear process. You may travel a long way toward your goal when, suddenly, a discovery on a side road will cause you to totally rethink your destination. I'll share examples of how this can happen later, but for now just know that you must be on the alert for such possibilities and you will need to be responsive to them. Bear in mind that many innovation initiatives are doomed because people, frequently those at the helm, are unable to acknowledge that new and unexpected information means their original target might not have been the best choice.

Successful innovation requires learning to be comfortable with the discomfort of not knowing where you are going. Depending on your nature, being uncertain about where you might end up is either the beauty or the horror of innovation. Some people enjoy getting into a car on a Sunday afternoon to just go exploring. The notion of getting lost doesn't panic them; in fact, they actually enjoy the challenge of it. For others, it's scary stuff to stray more than 50 miles from home without a road map prepared by the auto club, detailing every mile to be traveled.

2. The Lovers of the Status Quo Will Fight You Every Inch of the Way

As a champion of innovation, you will need to encourage people in your organization when they become squeamish at the prospect of traveling without a precise and unalterable road map. You will also have to deal with some folks who would prefer never to leave home at all.

Not everybody wants to move into the future. People who like things the way they are will feel threatened by any effort to create change. You will find barriers being thrown in your path by many of these fellow travelers as well as by the culture of the organization you are trying to drag into the future. Some of these obstacles will be inadvertent and easily removable, while others will be quite intentional and difficult to overcome.

This brings me to my favorite quote about innovation, which was forwarded to me many years ago by an appreciative client at Pitney Bowes named Ron Sansone. It sums up the difficulties you will face in dealing with those who desperately cling to the way things are:

> It must be considered that there is nothing more difficult to carry out, nor more doubtful of success nor dangerous to handle, than to initiate a new order of things. For the reformer has enemies in all those who profit by the old order, and only lukewarm defenders in all those who would profit by the new order, this lukewarmness arising partly from fear of their adversaries, who have the laws in their favor, and partly from the incredulity of mankind, who does not truly believe in anything new until they have had actual experience of it.
> —Niccolo Machiavelli, 1513

Sounds like a hard road to travel, doesn't it? Unfortunately, little has changed in the nearly 500 years since Machiavelli wrote that statement. This is why it's so important to understand as much as possible about the seemingly circuitous route that innovation takes *and* to be 100 percent committed to the objective of your trip. Don't start down the innovation highway without this knowledge and your total resolve because this journey is not for the uninformed or the faint-hearted.

3. How You View Your Competitive Position Affects Your Ability to Innovate

One of the reasons it is such a challenge for an organization to embrace change has to do with how it perceives its competitive position. That perception greatly impacts its ability to innovate.

Please note that I used the terms *perceives* and *perception*. This is because I find that most companies are amazingly naive about their competitors' activities. They are also surprisingly myopic about who their competitors really are. And, they are often wrong in assessing their relative competitive position in their markets. (For ideas on how to overcome the Competitive Myopia roadblock, see Chapter 11.) Nevertheless, because organizations tend to run almost entirely on their perceived realities, it is those perceptions of their competitive positions that tend to drive their innovation behaviors and practices.

I usually encounter companies positioned at the extreme ends of the spectrum of competitive perception. At one end are those who consider themselves to be market leaders, while those who feel beleaguered and frightened by their competitive vulnerabilities are at the other end. Each provides a different challenge (and opportunity) to the innovation consultant:

- The vibrant company is well aware that it must innovate to sustain its leadership. But, in its heart, it really wants innovation to be clean, nondisruptive, and, most of all, painless. Therefore, its innovations tend to be small steps such as line extensions, packaging improvements, new flavors or colors, and other minor, incremental differentiators.
- A company in trouble is far more receptive to attempting something new, but usually lacks the resources to support anything truly innovative. Again, deep down, it also doesn't really want to change. In *Management of the Absurd,* Richard Farson explains the challenge for the failing company this way:

  Deeply troubled companies don't usually seek help. And when they do, they have a hard time benefiting from it. The situation parallels one in psychotherapy. Psychotherapy is usually ineffective for severely mentally ill people; it works better for well people. The healthier you are psychologically, or the less you may seem to need to change, the more you can change.[1]

4. Innovation Requires a Passion for Paradox

Innovation is full of paradox, a word many people misunderstand and define incorrectly. My Webster's dictionary defines paradox as "a seemingly contradictory or absurd statement that expresses a possible truth." My earlier statement that those who pursue innovation need to be comfortable with discomfort is a paradoxical statement. The notion that the companies that most desperately need innovation are least likely to achieve it is yet another paradox of innovation. Struggling against or denying—rather than embracing—these and many other quixotic aspects of innovation is sure to lead to failure.

Again, Richard Farson explains the overpowering, instinctive desire to avoid dealing with paradox when he says, "Our natural inclination when confronted with paradoxes is to attempt to resolve them, to create the familiar out of the strange, to rationalize them."[2]

When new ideas surface that contradict old ideas that are dear to our hearts, we often lack the ability to juggle the two in our minds, so we end up resisting the new idea. Being unable to rationalize paradox, we push it from us, not understanding that one thing talented innovators have in common is an ability to contend with paradox.

My own love of paradox is why I chose to name our company Creative Realities. I wish I could say that I knew instinctively that we would constantly receive mail and phone calls for Creative Realties (as in real estate) but I didn't. However, the fact that we do get these calls just reinforces Farson's point: People automatically transform our name, which can be somewhat puzzling if you don't know what business we're in, into something that is more familiar—a real estate firm.

The desire for creative realities is another paradox of innovation. You need something new but you also need something that can be made real and marketable. Bringing those two concepts together is extremely difficult.

5. There Are Different Types (or Levels) of Innovation to Which You Can Aspire

There are three ascending tiers of innovation that you can pursue. The level you choose will dictate the simplicity or complexity of the journey you will undertake. Unfortunately, far too many companies expect to achieve the highest levels of innovation while only providing the tools and support for lower-level success, at best.

The three levels of innovation are:

1. *Incremental innovation.* This consists of small, yet meaningful improvements in your products, services, and other ways in which you do business. These tend to be the "new and improved" innovations we are all bombarded with every day: new flavors, shifts to better or all-natural ingredients, packaging improvements, faster/slower functioning, just-in-time supply chain enhancements, bigger/ smaller sizing, cost reductions, heavier/lighter weight. We see them every day and they help extend product, service, and business life cycles and improve profitability. They can be easily visualized and quickly communicated and give you something new with which to grab consumer attention in an increasingly noisy marketplace.

2. *Breakthrough innovation.* This is a meaningful change in the way you do business that gives consumers something demonstrably new (beyond "new and improved"). Breakthrough innovation produces a substantial competitive edge for a while, although the length of time anyone can maintain such an advantage is growing increasingly shorter. Examples would be: pasteurization, refrigerators, passenger jets, power steering, air-conditioning, clothes washers and dryers, the Sony Walkman, the minivan, fax machines, e-ticketing, frequent flyer programs, digital cameras, the Mobil SpeedPass, and Hertz #1 Gold service.

3. *Transformational innovation.* This is usually (but not always) the introduction of a technology that creates a new industry and transforms the way we live and work. This kind of innovation often eliminates existing industries or, at a minimum, totally transforms them. For this reason, transformational innovations tend to be championed by those who aren't wedded to an existing infrastructure. Examples include moveable type, the piano, gunpowder, electric lighting, telephones, automobiles, airplanes, computers, digital photography, and the Internet.

Now, before you go lobbing bazooka shells at my definitions and my examples, save your breath, because one person's incremental innovation is another person's breakthrough innovation. What I mean is that Ann's breakthrough innovation may be Harry's transformational innovation. The important thing is to understand that you can aspire to ascending levels of marketable newness. The bucket any one innovation fits into is directly related to how you define your world and each of us defines our world differently.

For instance, several years ago Vlasic Pickles launched Sandwich Stackers, a pickle cut lengthwise that could be removed from the jar and immediately placed in a sandwich. Because of its lengthwise cut (versus the then-available cross-cut pickle chips), the pickle fit better on the sandwich and provided more pickle in every bite.

Here's an example of an idea that could be located anywhere on the innovation spectrum (incremental, breakthrough, or transformational) depending on your perspective. The marketers (and probably most consumers) would say that a sandwich-sliced pickle was a no-brainer, incremental idea, because it made such common sense and could be easily visualized and communicated (which was true).

However, imagine that you are a manufacturing process engineer charged with the task of devising a way to mass-produce lengthwise-sliced, pickled cucumbers. Remember that, like snowflakes, no two cucumbers are exactly alike. For this engineer, you are challenging every fundamental in the manufacturing process. You are asking for breakthrough, even transformational, innovation.

I would argue that Vlasic's manufacturing colleagues would define Sandwich Stackers as breakthrough because it created a whole new segment in the pickle business and caused every one of Vlasic's competitors to scramble around to quickly launch their own versions. The point I'm trying to reinforce, though, is that one person's incremental, no-brainer innovation can often be another person's breakthrough, or higher. How you categorize something depends, if you're the innovator, on what your role is in making it happen or, if you're the consumer, on how significant an impact it has on your life.

Transformational innovation is exceedingly rare. Think about it: How many truly new-to-the-world ideas happen in a year? In a lifetime? Not many!

Yet, in some ways, transformational innovation is easier to pursue because the change required to achieve it usually doesn't rely on an existing entity that is committed to the old way of doing things. That's why we often find transformational innovation coming from start-up companies. But no company can survive by pursuing only transformational innovation. I can't think of a single organization that has succeeded over time in this mode alone.

Truthfully, most companies are content playing only in the incremental arena despite their protestations to the contrary. I was thrilled a few years ago, as the innovation wave began to build in the wake of reengineering, when I heard business leaders starting to declare that 25 percent of the products and services they would be marketing in five years didn't exist. Imagine my disappointment—and the deflation of their own people's expectations—when it became clear that in too many cases what they meant was merely adding aloe as an ingredient or offering a different color.

It's dangerous to set your sights only on incremental innovation. Many would argue that the term *incremental innovation* is an oxymoron. Nicholas Negroponte of MIT's Media Lab says that "incrementalism is innovation's worst enemy."[3] In business, I don't totally agree with him. What I do believe is that a strategy for pursuing only incrementalism is innovation's worst enemy.

I recommend that businesses take a portfolio approach to their innovation efforts. By that I mean the concurrent pursuit of at least two of the three levels of innovation. A portfolio of incremental and breakthrough innovations can provide most companies, in most industries, with an edge over their competitors, without letting them delude themselves about being focused on transformation.

Not that I want to discourage you from going for transformational innovation. Just recall what I said in Chapter 1 about trying to hit a home run all the time. Go ahead and aspire to be your generation's Hank Aaron or Babe Ruth but also realize the importance of having a high batting average. Don't be so focused on hitting one out of the park every time that you ignore perfectly good ideas that could drive in runs.

Putting all your efforts into only one of these types of innovation places you at risk. Many industries, like packaged goods, thrive on incrementalism. There is a place for it in most industries. But it can't be your only focus. Breakthrough innovation is expensive, messy, and uncomfortable. Nevertheless, it is becoming increasingly necessary to pursue if you want to remain relevant. Gambling on transformational innovation alone is too risky. You just can't afford to make any one type of innovation the sum total of your innovation strategy.

The focus of this book is on tools and techniques that will foster breakthrough and transformational innovation. The alterations in corporate mind-set and support mechanisms necessary to pursue these higher levels of innovation will be the most dramatic. Therefore, they are the easiest to define and illustrate (and, in some ways, to adopt). Incrementalists can randomly select from these offerings as needed to meet their goals. Should they decide in the future to broaden their objectives to include breakthrough innovation, they will already have surmounted some of the hurdles toward that goal.

6. Individual and Group Behavior Puts Numerous Barriers in the Path of Innovation

Mastering the people part is one of the critical keys to successful innovation. Therefore, you will be better prepared for your innovation journey if you know ahead of time what innovation roadblocks and speed bumps to expect along the way. Innovation speed bumps and roadblocks are the human behaviors that slow and even stop innovation from happening.

Speed bumps are behaviors and bad habits that are manifested by—and, therefore, are controllable by—individuals. In contrast, roadblocks are norms and behaviors that are group-based. Because roadblocks are cultural, they are harder to overcome than speed bumps caused by individuals. Both speed bumps and roadblocks can be fatal to innovation, so identifying and minimizing both is absolutely essential.

Innovation Speed Bumps Are Everywhere

The Bazooka Syndrome is a perfect example of a speed bump that exists in nearly every organization. A creative, new idea is extremely fragile. Like a seedling, it requires tender, loving care. Yet, something basic in human nature seems to prompt most of us to look first at the impossibilities of a new concept rather than its possibilities. We seem more adept at finding reasons to reject new ideas than at nurturing them. In short, we tend toward the bazooka as our first reaction to newness. And the humorous bazooka is the most damaging kind of bazooka because it is impossible to defend against humor.

Theoretically, speed bumps like the humorous bazooka should be avoidable through increased awareness and the introduction of skills focused on individual behaviors. But it's not that simple. Even people who have the proper training sometimes slip up. For example, I have never forgotten an incident, almost 16 years ago, when I used a bazooka on a friend at a party. After two years as a professional innovator, I should have known better.

The Night I Used My Bazooka on a Friend

In social situations, doctors I know tend to hear everyone's medical ills while my lawyer friends get asked for free legal advice. I get everyone's creative ideas! One night I ran into Steven at a friend's dinner party. Steven was in the shoe business, and it shouldn't have been surprising when he excitedly told me that he wanted to get my reaction to a name for a new boat shoe for kids. "We've got a great name. We're going to call them Undertoes."

Because it was a social event, and I didn't have my "work hat" on, my immediate response was, "Steven, do you want parents picturing their kids floating out to sea with your shoes on their feet!?" You can imagine the totally deflated look that appeared on his face.

Well, I went home and barely slept that night. I was so embarrassed. I had shot down my friend's idea with the first words out of my mouth. I actually woke up my wife, Amy, with my tossing and turning. "What's the big deal?" she said. "It wasn't a good name. Go to sleep!"

She didn't understand my distress. For two years, I had been teaching my clients not to do this. By taking a bazooka to his fragile idea, I had betrayed everything I stood for.

I called Steven first thing the next morning to apologize. A boat shoe for kids was a terrific idea. Maybe the name needed work, but Undertoes did have a strong, nautical theme. By aiming right at the flaw, I had squelched that candidate for his product's name. It never had a chance to grow.

If someone who spends his working life dedicated to nurturing innovative practices can make such a basic mistake, imagine how hard it is for people who haven't been trained in the skills that support innovation to avoid using their bazookas.

Speed bumps are relatively easy to identify. However, addressing them can have a powerful effect on an organization's ability to nurture and develop new ideas, helping to pave a smoother road to innovation. (See Chapter 8 for more on speed bumps and how to recognize and manage them.)

Innovation Roadblocks Are Even More Challenging

Unlike speed bumps, which can be addressed individually, innovation roadblocks can only be addressed through an organized set of techniques and skills that help manage group interactions. Roadblocks are also more difficult to eliminate than speed bumps because overcoming them requires cooperation, teamwork, and fundamental alterations in attitude and organizational behavior. Some of the roadblocks explored in this book are:

- *The Consumer Conundrum.* This involves understanding what it takes to be truly market-focused without falling into a pattern of following consumers instead of leading them.
- *Garbage In/Garbage Out.* This is the tendency to ideate feasible ideas and then use market research to justify them.

- *The Tyranny of Numbers.* This is when you avoid true risk-taking on potentially breakthrough ideas because they can never be numerically justified.
- *The Talent Assumption.* This is a cultural conviction that "some people have it and some don't" when it comes to creativity.
- *People Myopia.* This is when you too narrowly define the types of individuals who are selected to join an innovation team.
- *The Decision-Making Pendulum.* This is the struggle to balance autocracy and consensus on the team.
- *The Leadership-Empowerment Fable.* This is false empowerment and yo-yo decision making that results in the destructive "guess what's in my head" game.
- *Ready-Fire-Aim.* This is total absence of a strategic vision as the touchstone for decision making.
- *Information Overload.* This is the absurd assumption that lots of data automatically translates into insight.
- *Newness/Feasibility Schizophrenia.* This is the tendency to talk new, but walk feasible.

Innovation is never mistake-free. It can be painful, discouraging, expensive, time-consuming, and, worst of all, the outcome is always uncertain. All of these are characteristics that most people—and hence, most corporate cultures—don't support.

Most company cultures unintentionally encourage the construction of innovation roadblocks by creating a corporate ethos that punishes people for making mistakes.

I have a cartoon on my office "Wonder Wall" that shows a poor schlump sitting at a desk. One sign on the wall to the left of his desk proclaims, "THINK!" Another sign to the right of his desk says, "But don't get any ideas." What message does your organization send?

[i n n o v a t i o n   f u e l]

- Paradox lies at the heart of innovation, so embrace it. Don't automatically resist new concepts that run counter to ideas you've long held dear.

- Become comfortable with traveling toward an unknown destination and remain flexible along the way, always willing to take interesting detours that might prove fruitful.

- Develop strategies to cope with the people in your organization who fear change. Learn to recognize and overcome the barriers such folks put in your path, either as individuals (speed bumps) or as groups clinging to a corporate culture that is resistant to innovation (roadblocks). Be aware that fear of failure hides behind many of these barriers; work to develop a culture in which failure is not punished, so people can become comfortable with pursuing the unknown.

- Be constantly alert for instances where your own words or deeds cause an innovation speed bump. Eliminate idea bazookas, humorous or otherwise, from your behavior.

- Make sure the perception of your company's competitive position and your assumptions about the strengths and weaknesses of your competitors are realistic. Put yourself in their shoes and consider how you would respond to your company's innovation plans if you were the competition. Also, be certain you have correctly identified all possible sources of competitive threats, not merely the obvious ones.

- Do not put all your innovation eggs into one basket. Instead, take a portfolio approach to innovation that allows you to pursue at least two or even all three types of innovation—incremental, breakthrough, and transformational.

Notes

1. Richard Farson, *Management of the Absurd: Paradoxes in Leadership,* Simon & Schuster, New York, 1996, pg. 85.
2. Ibid., pg. 13.
3. Nicholas Negroponte, "The Balance of Trade of Ideas," *Wired,* April 1995.

CHAPTER THREE

Innovation Is
the Science and Art
of the Educated Gut

N ow get ready for some ideas that could really make your head spin—ideas about innovation that may sound like utter heresy.

Here are three not-so-obvious truths that challenge the philosophy underlying the approach to innovation taken in many, if not most, companies:

1. Real innovators know they are smarter about the future needs of consumers than consumers are.
2. Real innovators will ignore almost any evidence—no matter how logical—that doesn't support their vision because they believe so passionately in their ideas.
3. To be truly innovative, you must take a quantum leap into the future, where the real opportunities for breakthroughs exist *and the future can't be quantified!*

These ideas may sound contrary to almost everything your organization has done up to now to support innovation. For example, if your business requires extensive, quantitative support for launching anything new, you will find the ideas in this chapter extremely difficult to accept. So, be wary of spreading these views. It can be a very tough job within some companies. But it is a job worth accepting if you want to be part of an organization that is continuously successful at innovation.

Let's look at these ideas one by one and determine together why they make complete sense to embrace if you are committed to pursuing breakthrough or transformational innovation.

The Consumer Conundrum

Real innovators know they are smarter about the future needs
of consumers[1] than consumers are.

As you undoubtedly know, being market-driven, consumer-focused, or customer-centered has become a mantra in most companies. Obviously, how well this important goal is satisfied varies dramatically from company to company. I believe many fail to achieve this objective because they misinterpret its meaning. They think that being market-focused/centered/driven (take your pick) means giving consumers what they *say* they want. This is an invalid assumption that might well cost you your business.

Following the market, rather than leading it, is a major innovation roadblock. The real truth—and the challenge for business innovators—is in identifying how to lead consumers, not follow them. This is the consumer conundrum hurdle you must overcome if your company is to be the winner in a business environment that grows more competitive daily.

Really satisfying consumers, by leaping beyond them, requires that you develop skills to read these people better than they understand themselves, then link marketplace desires with your knowledge of available technologies and capabilities to deliver innovations they haven't even envisioned yet.

For example, consider Chrysler's decision to add a second sliding passenger door on the driver's side of its minivans. Market research over a period of years indicated little consumer interest in this concept. Only about 30 percent of minivan buyers in 1995 rated this option as desirable, while 50 percent said they weren't at all interested.

Despite this less-than-enthusiastic response from consumers, Chrysler went ahead and added this extra door to both its Chrysler Town & Country and its Plymouth Voyager models, overcoming some big production obstacles to do so. The result: almost 90 percent of their buyers opted for the feature, even though the extra door cost $600 or more on some models. The rest of the auto manufacturers, who may have abandoned a similar option because of their consumer research, were forced to play catch-up.

If your innovation efforts are grounded solely in your understanding of where your consumers are today, you will not succeed. By the time you get your innovations launched, the marketplace will have already progressed well beyond where it is today.

The lead time to market launch is a critical factor in the idea development equation. During that time, marketplace factors will evolve and the consumer will change. The goal of the innovator must be to stay *ahead* of the marketplace—to forecast wants and needs that may not yet have been articulated and determine ways to satisfy those emerging needs creatively.

The Gretzky Paradigm

Here's a simple statement that captures this concept. Legend has it that when asked the reason for his phenomenal success, hockey superstar Wayne Gretzky responded, "I don't skate to where the puck is. I skate to where it's going to be."

[**FIGURE 3.1**] **The Gretzky Paradigm**

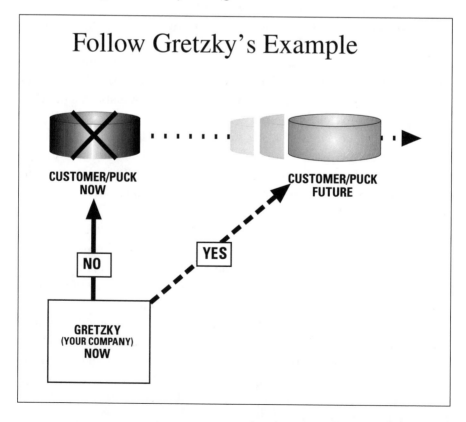

Personally, I don't really care if it was Wayne Gretzky (or as some believe, his father, Walter) who spoke those words, because it's the mental paradigm it suggests to innovators that interests me. I know there's something to it because every time I relate this quote to a client team, they get it immediately. The paradigm is so elegantly simple that little else needs to be said in explanation.

And I always know I've struck a resounding chord (like with the bazooka metaphor) when members of the client team begin to share stories that support the Gretzky Paradigm. A perfect example of this occurred a couple of years ago when I shared the paradigm with a client team at one of America's leading banks.

In the early 1990s, before PC banking became commonplace, this bank presented a fairly advanced concept for PC banking to a cross section of its small business consumers. This segment is particularly important in banking because small business owners tend to be highly entrepreneurial. Therefore, they often require lines of credit. They also tend to keep both their business and personal accounts with one bank.

The folks at our client bank had concluded, as the result of some earlier market research and their own good instincts, that this might be a worthy, market-driven innovation to develop and launch.

Unfortunately, in some follow-up concept research the bank's small business consumers rejected the notion of PC banking out of hand as being unnecessary for their needs. So, our client didn't pursue the idea.

Three years later, those same consumers were clamoring for PC banking, which our client's competitors had launched quite successfully. Those same small business consumers who had initially rejected the PC banking concept were frustrated that the bank wasn't offering it when they wanted it. Our client forfeited three years of developmental lead-time and was losing customers from a particularly profitable segment of their business.

The bank thought it was being market-driven. Chrysler would probably argue that it was, too, when it added the driver-side van door. But Chrysler had solved the consumer conundrum and, in this instance, the bank hadn't. I will revisit this conundrum and introduce ways to address it in Chapter 10.

You Can't Quantify a Beginning Idea

Real innovators will ignore almost any evidence—no matter
how logical—that doesn't support their vision because
they believe so passionately in their ideas.

To be truly innovative, you must take a quantum leap into the future,
where the real opportunity for breakthroughs exists
and the future can't be quantified!

These two ideas stand in opposition to the absurd but amazingly common demand that innovation be made quantifiable so that failures will be avoided. Real innovators know that when it comes to breakthrough innovation you just have to trust your educated gut (another paradox).

Not many companies understand this concept of educated gut nor are they comfortable with its approach. Instead, they try to take the risk-free route to innovation. Of course, innovation is never risk-free, so the next thing you know, these companies encounter head-on collisions with two closely linked and incredibly lethal roadblocks:

1. Garbage In/Garbage Out, and
2. The Tyranny of Numbers.

The Garbage In/Garbage Out Roadblock

There is a difference between just following gut instinct and applying an educated gut to a problem. The educated gut is the result of highly empirical exploration and open-mindedness. I believe it's one of the characteristics that sets the great thinkers apart from everybody else.

To begin to understand this critical difference, here's how innovation by gut instinct usually plays out. You work for the most profitable division of a multinational conglomerate. A major competitor is beginning to threaten your market position so senior management has challenged your group to come up with some exciting new ideas. Your team jumps right into ideation without doing any preliminary research on any of the four factors that drive innovation, which are: consumers, technology, competition, and regulation. Your team figures it knows its market well enough to develop some good ideas; after all, the team members live in it every day.

In a brainstorming session, the volume of ideas is high but the quality of those ideas is low. The organization finds itself having to choose from an array of ideas, most of which have a gee-haven't-I-heard-this-before? sound to them. But, needless to say, this doesn't prevent the innovation effort from moving forward.

Nobody has the nerve to say, "Hey, these ideas are all old. We should start over and do this differently." Competitive pressures, along with the burden on individuals to quickly produce the results management wants, don't allow for anything except driving the innovation effort ahead. In the end, what comes out the other side—the actual end product or service—is of the same quality as what went into the effort to begin with: Garbage In/Garbage Out.

This roadblock arises for two primary reasons:

1. We ideate too early in the innovation process (because we don't know any other way) and devote the majority of our innovation time line to qualitative testing and quantitative assessment steps (with which we are comfortable).
2. We don't prepare our minds properly for ideation.

Why do we jump into idea creation without properly preparing our minds with the fresh food for thought needed for successful ideation? In part, because we're just too darned excited about having the chance to break out of the strictures that are so much a part of normal business life!

In almost every aspect of business management, our thoughts and decisions are constrained by rules, systems, history, and lots of other parameters imposed by the infrastructure. People working within these environments are victims of what my partner, Cris Goldsmith, calls "the electric fence syndrome."

This is a phenomenon Cris observed when a farmer informed him that the electrified fence encircling his field had been shut off for years, ever since the first few cows were zapped. After a couple of jolts, the herd of cows institutionalized a stay-away-from-the-fence policy. People operating in groups tend to behave much like the farmer's cows. We don't test the boundaries because we've seen people get zapped for trying to work creatively outside the norms.

Contrast this behavior with what we expect will happen when we're asked to participate in ideation of any sort. We are given a rare opportunity to liberate our brains from their daily constraints. We are encouraged to think outside the box or think beyond the nine dots.

Having been given this freedom, we are loathe to submit to any rigorous analysis of anything before ideating. Our goal is to think creatively. So, once we know what our objective is, we just let the creative ideas flow fast and furiously.

Unfortunately, in most instances, we haven't been trained in any creative problem-solving skills. We also haven't done the prework that would bring us to the ideation session with minds filled with notions of where the market might be headed, how technology might be used in new and different ways, what competitors might be doing that needs to be counterbalanced, or what's on the regulatory horizon. Is it any wonder that the ideas we develop are lackluster and likely to fall short of our innovation objective?

The Tyranny of Numbers Roadblock

Having committed to the implementation of a Garbage In/Garbage Out idea, an organization tries to justify its support by expending a great deal of time and money on quantitative research to "prove" the idea will work in the marketplace. This kind of research is, in fact, where the greatest effort is spent in most corporate innovation efforts. In reality, this is time and a lot of money poorly spent.

We like to play where we are comfortable and we are comfortable manipulating numbers. We are uneasy about trusting our instincts and intuition. Therefore, we try to quantify fledgling, beginning ideas far too prematurely in the innovation process. It doesn't work! The challenge to any company aspiring to be innovative is to break free from the mistaken belief that innovation can be quantified.

If you prefer to cling to the notion that you can make innovation foolproof if you just do enough market research, let me point out something that has long puzzled me about the consumer packaged goods and consumer service worlds. Statisticians tell us that, at best, only one or two out of every ten new products and new services that are launched in the marketplace are considered a success. Yet, the majority of these products and services are subjected to some form of quantitative testing before being launched. If so many of them fail, what's the value of these quantitative testing rituals?

Why do companies continue to devote so little time and money to the strategic and pre-ideation front end of innovation while spending so much time and money quantifying the back end? Maybe that's why it is so often referred to as the fuzzy front end of innovation.

In organizations where innovation lives or dies by market research numbers, it would be absurd to advocate an idea based primarily on gut instinct. Ideas that don't have a thick report confirming their appeal to consumers go nowhere in such cultures. Nobody is willing to climb out on a limb to support something that feels right but doesn't have reams of supporting data.

The unsuccessful, consumer-following innovators abdicate intuition in favor of defendability. In a consumer-following company, if research demonstrates that consumers want X, and you give them X, your decision is quantifiable and is always defendable. And, as we all know, in risk-averse cultures quantitative documentation for a hypothesis is critical to gaining support for moving an idea forward. People working in such cultures know they can't make decisions based on anything that can't be proven by the numbers.

In business cultures that punish mistakes, using qualitative measures and gut feel to forecast where the marketplace will be when new ideas launch is dangerous because such nonquantifiable forecasts are virtually indefensible. In such cultures, adopting the Gretzky Paradigm will fail because there are no mechanisms to gain the support of the decision makers.

This mentality, which causes everyone to feel the need for a cover-your-ass strategy, leads many innovation teams to take the safe road, to wimp out, and ultimately fail. In such an environment, when a new product bombs, you can always survive with a shrug and say, "Hey, we gave them what they said they wanted. I've got the research data right here to prove it." And your job will be secure.

The Tyranny of Numbers roadblock is particularly destructive if the new idea is, potentially, really big, because the bigger the idea, the higher up it must travel through the decision-making labyrinth. This will virtually guarantee that a potentially breakthrough idea will never see the light of day. I will discuss this further in Chapter 6.

What True Innovators Must Do

True innovators—who lead rather than follow consumers—use market data as only one element of the decision-making mix. True innovators trust their educated gut above all else.

True innovators are gamblers. The ability to predict where the marketplace might be headed two or three years down the road is both the art and the risk-orientation needed to achieve successful innovation. By thoroughly accepting both of these labels—that innovation is an art *and* a gamble—organizations can best equip themselves to achieve breakthrough innovation. This requires acceptance of a gambler's mentality. As Swiss mathematician Euler said, "Science is what you do after you guess well." In business innovation, I believe this—passionately!

The successful gambler knows that risk and acceptable losses are part of the game. Similarly, a successful innovator knows that staying ahead of the marketplace is where the real rewards of innovation lie, as do the real risks.

But risk-taking is expensive both in financial as well as human resource terms. In response, many companies have begun to do what they do best: creating internal, highly visible innovation processes. As I discussed earlier, the creation of rigid innovation processes ignores the need for flexibility and the all-important people part of innovation.

Also, no process will ever foster real innovation unless the organization is willing to live by its best-educated (usually nondefensible) assumptions. Otherwise, it will only reinforce and institutionalize those cultural roadblocks that stifle real innovation. If you can't find a way around these innovation roadblocks—Garbage In/Garbage Out and the Tyranny of Numbers—you'll be forever stuck at a red light on the innovation highway.

How to Overcome These Two Roadblocks

The Garbage In/Garage Out and the Tyranny of Numbers roadblocks are, obviously, closely linked. Where you see one, you usually see the other. Establishing an attitude, which flows from the top, that risk-taking is okay, even encouraged, is a critical step to minimizing both of these barriers. Leadership must champion the philosophy of Elbert Hubbard, a prolific author and a leader of the early-1900s' Arts and Crafts movement, who said, "Constant effort and frequent mistakes are the stepping stones of genius."[2]

Equally important, people at all levels of the organization must be truly empowered so they can feel safe enough to trust their intuition instead of relying on expensive, time-consuming, and probably futile research. This requires the establishment of a clearly articulated decision-making process guided from the top. Again, Chapter 6 addresses the decision-making structure needed to meet this part of the innovation challenge.

The Well-Prepared Mind (The Magic Ingredient for the Educated Gut)

One of the most important tools for overcoming the Garbage In/ Garbage Out roadblock requires developing an appreciation for the well-prepared mind that I mentioned in Chapter 1. This is an extremely important concept that is critical for the support of breakthrough innovation. My enthusiasm for the power of well-prepared minds is rooted in two related beliefs:

1. Most people need to learn (relearn, actually) how to think creatively before embarking upon an innovation initiative.
2. Breakthrough ideas nearly always appear to be flawed at first so people also need to be trained in ways to evaluate ideas without killing them.

What makes great thinkers truly great is not just their ability to come up with creative ideas but also their facility for detecting the simple brilliance in seemingly absurd statements. It's what Richard Farson calls, "the invisible obvious."[3] The Edisons and Einsteins of this world appear to be born with this talent but it can also be taught and facilitated.

An Excursion into the Absurd

Let me give a real example that demonstrates how critically important it can be to provide training in both creativity and idea-evaluation at the start of any innovation effort. The story also illustrates what can happen when leaders are willing to be role models for risk-taking.

One of my oldest client relationships is with Rich Products Corporation in Buffalo, New York. It is the largest family-owned frozen food manufacturer in the United States, and sells products primarily through food service outlets (restaurants, schools, hotels, etc.) and in-store bakeries. You may be familiar with some of Rich's forays into the supermarket: Coffee Rich and Farm Rich nondairy coffee creamers and Rich's eclairs, to name a couple.

Some years ago, I captained a team that facilitated a strategic innovation program for Rich's, focused on the creation of new product ideas. At the beginning of the project, we did some training to teach the group the two key skills that support breakthrough thinking: connection making and open-minded evaluation (these will be discussed in greater detail in Chapter 9). Then, before ideation began, some exploratory conversations were held with restaurant operators and other providers of food prepared away from home (which is how they describe their business). This helped prepare people's minds for ideation.

To further help stimulate creative thinking, the client invited my friend Barbara Caplan from Yankelovich Partners to present some thought-stretching trends and marketplace insights to stimulate the client's invention team throughout the brainstorming session.

One of the many skills that every well-trained facilitator develops involves techniques to take the group on a controlled mental flight of fancy. These techniques are called excursions, or side steps, or managed serendipity. During the opening-up, wishing part of the brainstorming session, I decided to take the group on one of these trips to stretch the members' thinking and thrust their mental feet into midair.

First, I instructed everyone to forget about Rich Products and think of a memorable scene from a favorite movie. You can imagine how surprised everyone was to hear this instruction, even though they had gone through training in ideation skills. After a slow start, everyone offered up a favorite scene.

Somewhere in this exercise, someone from the group said his favorite scene was the one in *Animal House* where John Belushi stuffs his face with food, then smacks his cheeks with his hands spewing food all over everyone while declaring that he's a zit. In all, I collected almost two dozen scenes from different movies.

Next, I told everyone to pick one of these movie scenes (preferably *not* the one they had offered), then take a clean sheet of paper and use it to doodle and make mental connections using that movie scene. After about three minutes of total silence and frantic scribbling, I challenged them to use their scribbles and doodles about the movie scene to generate an absurd wish for a new product idea for Rich's. I encouraged them to come up with an idea that would get them fired if they offered it up outside that room.

The group was a bit nervous but intrigued. Again, after a slow start, one member of the group offered up the following wish: "I wish we could make a bagged whipped cream that looks like a zit."

While I, of course, can't remember his exact words, our session participant described the thought process he had gone through to get to this truly absurd wish in roughly this way:

> Part of the idea came from my fascination with the movies and my wonder at how many times the movie crew might have had to shoot that scene . . . That led me to think about alternatives to mashed potatoes, which led me to whipped cream . . . Whipped cream led me to the operator conversations I recently participated in, where the operators shared their frustration with the messiness and unreliability of whipped toppings . . . That led me to the geometry of a zit and how perfect it is for dispensing.

I loved this explanation because it was an example of pure, free-flowing, unrestrained connection-making by a well-prepared mind.

Now, I want you to mentally picture yourself in that room. It's four hours later and that zit wish was number 136 out of over 300 wishes written on big sheets of paper hung around the room. You are handed a colored marker and asked to select the five most promising wishes on the wall for development in breakout groups the next day.

Would you have put a check mark next to wish number 136? Probably not, because our tendency is to select ideas for their feasibility, not their newness (Chapter 13 explains Newness/Feasibility Schizophrenia). This group was no different. Only two people put a check mark next to number 136 but one of them was the group leader.

That leader—who had been through a couple of these sessions with me and was more at ease with risk-taking—asked one of the breakout groups to play with that idea because the geometry of a zit tickled his fancy though he wasn't exactly sure why.

The concept that emerged from that breakout session was for a cone-shaped bag of whipped topping, with a nozzle embedded in the tip that would drop into place via gravity after the plastic point is sliced open. The entire contents of the bag would be dispensed directly from the bag, which could be easily discarded when empty. No preparation or cleanup mess. Their initial, very rudimentary drawing looked something like the diagram in Figure 3.2.

There's more to the story. Robert E. Rich, Sr. is the founder of Rich Products. Mr. Rich is the classic entrepreneur who has almost always been in tune with his customers' needs and has always trusted his educated gut.

[FIGURE 3.2] Bagged Whipped Topping That Looks Like a Zit

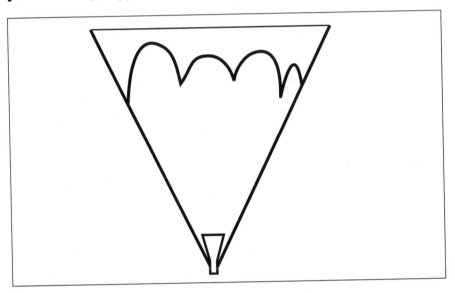

Shortly before our innovation project began, it seems that Mr. Rich had just returned from a trip to Japan, extremely enthused about some of the new packaging technologies he had experienced in use there. One of those technologies offered capabilities for heat-sealing fragile food products in clear plastic. When our invention team's concept encountered Mr. Rich's packaging technology in the lab, the result was a breakthrough new product called On Top nondairy dessert topping.

On Top won first prize in *Prepared Foods* magazine's tenth anniversary new products contest for innovation in the frozen food business the year it was launched, and it is now one of Rich Products' most successful product platforms. Since its launch, they have introduced chocolate and sugar-free versions, and they are exploring line extensions into some nontraditional arenas.

Obviously, I am very proud to relate this story because it celebrates a truly innovative client and it encapsulates so many of the truths we are exploring together in this book. This is how innovative teams can work together. Also, it shows how you can begin to overcome the Garbage In/Garage Out roadblock by creating an atmosphere where people feel free to take risks and by providing training that encourages people to delve fully into the ideation process. As a result, the Rich Products team was willing to embrace an absurd idea and see where it would lead.

Finally, this story also begins to point toward a new approach to innovation, which we will explore next in Chapter 4. It eliminates many of the causes of the Garbage In/Garbage Out and the Tyranny of Numbers roadblocks. By altering the flow of innovation, shifting the focus of discovery and exploration to earlier stages from the later stages, you can significantly improve the quantity and quality of ideas generated. Also, in many cases, you can shorten your time to market by eliminating the compulsion to perform endless research and quantification that drives the Tyranny of Numbers and results in Garbage In/Garbage Out.

[i n n o v a t i o n   f u e l]

- Accept that real innovators have a better understanding of their consumers' future needs than the consumers themselves do. Don't let being market-driven become an excuse for following instead of leading consumers.

- Follow the Gretzky Paradigm and develop processes that enable your company to anticipate consumers' as-yet-undefined desires. Combine these ideas with your knowledge of available and emerging technologies and capabilities to deliver innovations that consumers haven't thought of yet.

- Recognize the vast difference between following your gut instinct and applying an educated gut to a problem.

- Eliminate the Garbage In/Garbage Out roadblock by developing a team of well-prepared minds before embarking on idea creation. Do the necessary homework in advance of brainstorming instead of assuming you already know everything you need to know about your consumers and your marketplace.

- Don't kill beginning ideas by trying to quantify their potential too soon. Avoid the Tyranny of Numbers roadblock by accepting that, because the future is not a predictable place, innovation is not quantifiable. Build a culture where good ideas do not have to be defended with mountains of research before they are embraced.

- Accept that risk is an inherent part of innovation and allow people to take risks without being punished if things don't work out as hoped.

- Encourage people to champion ideas because the concepts feel right to their educated guts. Model this behavior by being a passionate idea champion yourself.

Notes

1. Just to clarify: In my lexicon, the customer is the channel intermediary, often a distributor or wholesaler, who becomes a critical distribution gatekeeper to the consumer. The consumer is the ultimate end user. Too many companies, especially business-to-business firms, place far too much emphasis on their channel customers, with whom they interact constantly, while virtually ignoring the end user who they rarely encounter in their normal course of business. This is a serious mistake because the consumer, the end user, is the person we all ultimately serve.
2. Todd Siler, *Think Like a Genius,* Bantam Books, New York, 1996, pg. 96.
3. Ibid., pg. 26.

CHAPTER FOUR

A New Flow
of Innovation

As I said in Chapter 1, no single fixed-process road map for innovation exists. Many of the companies who contact my firm are seeking an innovation process—a set of linear flowcharts in a book with clearly identified decision gates that they can internalize to achieve guaranteed results. They are puzzled and disappointed when we tell them that, when it comes to innovation, no single process will work anywhere, all the time. This does not mean that innovation should not have a programmatic approach; it will just be different for each effort.

Anyone who tries to convince you that he or she has a process that is guaranteed to work in *any* organization at *any* time is simply wrong, even if he or she promises one that will be customized to your organization. This is especially true if that process can be defined in a procedures manual. I've met so many CEOs who were terribly frustrated after spending up to, or even over, a million dollars for an innovation process in a fancy binder that ended up gathering dust on everyone's office credenza. What a waste! But, shame on anyone who thinks there is a one-size-fits-all method for innovating.

Remember, as explained in Chapter 2, there are three different types of innovation. Each requires a different level of sophistication in process. If this is true, then how could one road map address all situations?

Most critically, an innovation process in a binder overlooks the two keys for true innovation:

1. *The people part.* This is the successful interaction among people at all levels of your organization throughout what is often a lengthy, difficult undertaking. Basic human nature dictates that people can rarely be prevailed upon to fall into lockstep with a binder-delivered process that doesn't take into account your organization's unique culture.

2. *The need for flexibility.* You need flexibility by the bucketful. Remember this rule: Don't drive onto the innovation highway without a full tank of flexibility.

By its very nature innovation happens differently every time. One might even argue that the term *innovation process* is itself an oxymoron. This would be true if you define process conventionally.

When most of us consider business processes, we tend to think about fairly rigid, operational procedures for getting something done like a manufacturing process. Once a company has developed a successful way of accomplishing something, everything in the organization—the infrastructure, the measurement and reward systems, and, most of all, the people—begin to conform to and support what has historically worked. And, while there's always room for improvement, the formality of this type of process is key to its value.

Companies become very attached to processes that work and, as a result, tend to believe that the processes for developing new initiatives should look and function exactly like their existing business processes. With established lines of business, people get used to certain protocols, decision-making practices, production procedures, and profit profiles. Instinctively, they want to hold any new opportunities to those same, comfortable standards that they are used to. It doesn't work!

There's no denying that formal, repeatable processes are essential for running an established line of business. But, as a former colleague liked to say, "The *good* thing about business systems is that they work. The *bad* thing about business systems is that they work!" As these systems reinforce what already exists, they also create huge resistance to change. Without some counterbalancing techniques for nurturing innovation, they produce a corporate culture that chokes off efforts to produce newness rather than promoting them. This causes the that's-how-we-do-things-around-here mentality that dominates many workplaces.

The Paradox of Business Innovation

The paradoxical challenge to the future of any thriving business is to keep doing well those things that make an organization great while nurturing new possibilities, even those that might threaten the very foundation of that same business. The key is to understand that you must do both. In *Built to Last,* James Collins and Jerry Porras visualize this seemingly conflicting duality of purpose by using a simple permutation of the traditional Chinese yin/yang symbol with the following ideas: preserve the core, and stimulate progress.[1]

Stimulating Progress Requires New Tools

To address this paradox of needing to maintain the systems that keep your existing business successful while simultaneously supporting change, requires a much different approach than most companies take when embarking on an innovation effort. Top executives like to be viewed as clearly supporting the concept of change. They set lofty goals for innovations that will account for a large percentage of their business every five years. But, rarely do they provide for or, more importantly, sponsor and champion the painful changes in systems and cultural attitudes that real innovation demands.

Instead, organizations typically try to stimulate innovation with the same process mentality by which they manage their established businesses. This causes all sorts of problems. The paradoxical goals for managing an established business versus the pursuit of new ideas (formality versus flexibility; risk aversion versus risk orientation) produce a kind of organizational schizophrenia because people are being asked to simultaneously own what is while also embracing what could be.

You Must Be Flexible, Flexible, Flexible

What you need to do instead of clinging to a traditional business process mentality is to evolve a flexible set of innovation tools, to determine what will work in your particular organization at a precise point in time for the specific objective you have in mind. Again, here is where an intrapreneurial approach is helpful. Thinking and acting more like an entrepreneur than like a supporter of the status quo helps loosen the ties that bind you to the existing processes.

I believe there is a flexible set of procedural tools and a flow of developmental phases through which every innovation initiative must pass. I know and have repeatedly proven that it can help any team increase its chances of achieving success in the pursuit of innovation. I also call it an innovation process, but bear in mind throughout the rest of this book that, because we're talking about innovation, my definition of process is much different from the one to which you are accustomed. When I say *innovation process,* I am referring to an extremely flexible, people-focused, set of innovation phases and behavioral techniques.

This lesson is so important that I'm going to repeat it again: As you pursue innovation, don't waste your time searching for a formal, repeatable process like those you use to run your established businesses. Don't waste your time assembling a cross-functional team that reports, inefficiently, through the usual tangled web of decision making. Instead, consider the idea of having a flexible innovation process: a set of behaviors and practices that can be modified selectively to help any innovation effort.

The Five Phases of Innovation

Nobody can hand you a single road map with your path to innovation marked with a yellow highlighter. Instead, what I am going to give you is an outline of the phases of innovation and a new approach to how you can move through these phases in a way that will help you overcome major roadblocks. And, while I wish I could create a new term to describe this approach, so far nothing has worked better than flexible innovation process.

As a result of many years of collaborative teamwork and development by everyone who has been a part of Creative Realities since we opened our doors in 1988, we as a group have identified five phases that successful innovation efforts must pass through. As a tribute to all of the colleagues who have contributed to developing this model, I will often use the term *we* throughout the remainder of this book. Please understand that we means everyone who is now or has ever been employed by Creative Realities.

Think of this flexible process as the innovation landscape you will view from your vehicle as you travel your own innovation highway, because every innovation journey is different. These phases have a logic to them, but it is not a rigid logic. I think of their flow as the gestalt of innovation that we've recognized from watching numerous organizations' attempts to develop their own innovation processes.

The five phases of our flexible innovation process are:

1. *Setting Objectives: Determining the goals, touchstones and measures for your journey.* When you can align the goals for your innovation initiative with a clearly articulated, well-defined, strategic vision for your overall business, the means for achieving success become clearer and you have decision-making touchstones for the difficult days that lie ahead. Tangibility and measurability also increase.

2. *Discovery: Exploring the drivers of innovation.* You will begin this phase by developing John Saxon's well-prepared mind. This will require learning some new skills for creative ideation, problem solving, and teamwork. Armed with these skills, you will then creatively explore the four stimulators of innovation: technology, competition, regulation, and, most importantly, your consumer. You will expand the possibilities of what you might do, avoid competitive myopia, and be inspired to anticipate your consumers' unmet needs before even they can articulate them.

3. *Invention: Generating an exciting array of beginning ideas.* Keeping an eye on your objectives from Phase 1, and using the skills learned and discoveries made in Phase 2 to stimulate your well-prepared minds, you will be more likely to create truly novel, innovative concepts. These might range from easy, just-do-it ideas that can be immediately implemented with no muss, no fuss to potentially breakthrough ideas that will cause your head to spin because they are extremely exciting yet will be tremendously difficult to successfully develop and implement.

4. *Greenhouse: Letting your ideas grow toward reality.* You enter the Dark Night of the Innovator as you try to protect and nourish your nascent ideas. Nurturing and strengthening concepts and building in needed feasibility are this phase's goals. Here you will find answers that take a concept from "Gee, that's impossible" to "Wow, it could work!"

5. *Implementation and Launch: Actually getting your ideas to reality.* Navigating the hurdles to implementation can be the toughest part of innovation. Keeping everyone on board, making sure you get the resources you need, managing unexpected internal and external changes, and evaluating the results of implementation efforts so that

any necessary course corrections can be made quickly are all part of your work in this final on-ramp to reality.

Moving through the Five Phases

In each phase, you will need to select the innovation tools and techniques from a variety of sources that will work best for your objectives and with your culture. But bear in mind that any innovation process will fail without flexibility and good management of the critical people parts.

As I've already said, how you move through these phases will differ from company to company. It will also differ from innovation effort to innovation effort depending on which level of innovation you are pursuing. This process differs considerably from the one I described in Chapter 3, the traditional process used in companies that suffer from the Garbage In/Garbage Out and the Tyranny of Numbers roadblocks. The innovation process in those companies looks like the diagram in Figure 4.1.

[FIGURE 4.1] The Traditional Innovation Process

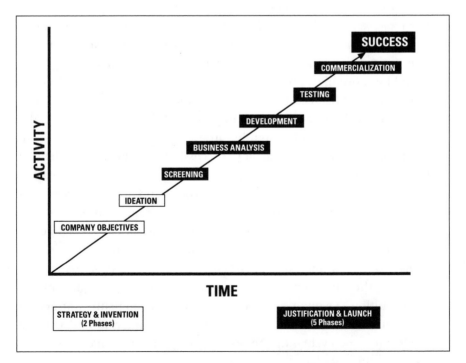

[FIGURE 4.2] The New Innovation Model

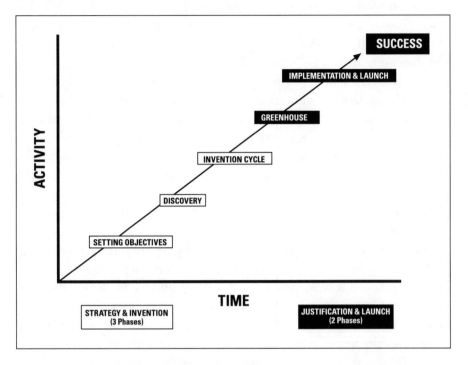

In contrast, Figure 4.2 shows how the five-phased innovation process we use looks if it is plotted along the same axes as the traditional model.

Our new model devotes far more time to the strategy and invention phases than has typically been the pattern in most organizations. In the next few chapters, you will learn why it is so important to spend a much larger portion of time in the Setting Objectives, Discovery, and Invention phases than has been done historically to build the knowledge base that then inspires the development of truly exciting, breakthrough ideas.

This change of emphasis can do two amazing things: It can speed up your innovation process, and, more importantly, it can also significantly boost your chances of success. It can achieve both these goals because it addresses those two killer roadblocks—Garbage In/Garbage Out and the Tyranny of Numbers.

This new approach galvanizes the organization, the whole organization, behind the creation and implementation of new ideas. It minimizes the need for extensive quantitative testing because, as already discussed, quantitative testing doesn't achieve all that it promises.

More importantly, people who have participated in an effectively managed innovation process are more willing to trust their instincts because their ideas are grounded in an expanded, strategically based vision and reality that includes considerable up-front input from the marketplace (not just the consumer).

With this new model of innovation, the final two phases (Greenhouse and Implementation/Launch) will proceed more smoothly and faster, and the outcome is likely to be more successful. By rethinking where you put your innovation emphasis in terms of time and resources, you can counteract the Tyranny of Numbers and the Garbage In/Garbage Out roadblocks.

Is Speed Important?

For a number of reasons, the time it takes to travel through the phases of innovation does vary. One factor that sometimes determines the speed at which you can travel is the level of innovation you are seeking—incremental, breakthrough, or transformational. In most cases, the greater the newness you are trying to achieve, the longer it will take to create and launch the change. But, as I will explain, this is not always so.

Another factor is your skill at removing and overcoming the innovation speed bumps and roadblocks that will invariably block your path. If you are very effective at this, innovations can be created and launched much more quickly.

This raises another speed-related determinant: the nature of the industry in which you operate. Each industry has a different cycle time for innovation. Regulated industries like banking, pharmaceuticals, and some forms of communications (like the postal system) tend to be far slower in their innovation cycle times. Consumer packaged goods and service businesses thrive on incremental innovation and tend to be in the midrange on cycle time (with food usually being the slowest). Cosmetics, fashion, and the high-technology arenas move at bewilderingly high speeds as does most business-to-business.

I mentioned above that cycle time is usually, but not always, linked to the level of innovation being pursued. High tech is the industry that is not as constrained as others. In fact, it is the only industry I've ever seen that gets away with throwing largely untested ideas into the marketplace, knowing in advance that additional revenues will be generated by introducing new, debugged versions almost immediately. With the emergence of the

Internet, this trend has become the norm for dot-com companies. And, for some reason, we all tolerate it.

Getting Started

Enough of the philosophy and process diagrams! Let's get going. First, I'll talk about how to assemble your innovation team and how this team can best be managed to help ensure success. Then, the rest of this book will describe the types of activities you can undertake during each phase of innovation and will show you how to overcome the speed bumps and roadblocks that typically arise along the way. It will help you develop a pre-emptive innovation strategy and significantly increase the odds of your succeeding at breakthrough innovation, a worthy challenge.

Please note that I said "significantly increase the odds." That's because I need to reinforce that failure is an integral part of innovation. There are no guarantees in the pursuit of breakthrough innovation. There are only ways of doing it better. But even your innovation failures can provide important lessons for future efforts if you truly believe that every experience teaches you something (as showcased in the Thomas Edison story in Chapter 1). As Samuel Beckett said: "Ever tried? Ever failed? No matter. Try again. Fail again. Fail better."

[i n n o v a t i o n   f u e l]

- Match the level of innovation you are pursuing (incremental, break-through, transformational) with the appropriate time, resources, and levels of decision making you will need to achieve your goal. Thus, because there is no single right way to innovate, flexibility is one of the most important ingredients in any innovation initiative.

- Embrace the paradox of business innovation by continuing to do well those things that are working right while simultaneously nurturing new possibilities, including those that might threaten your existing business.

- Avoid trying to apply your established, repeatable business processes to your pursuit of innovation. They will only constrict your ability to pursue newness and will end up creating frustration that makes people unwilling to participate in future innovation efforts. Instead, develop new ways of working in your pursuit of innovation.

- Apply a flexible innovation process that has five phases: Setting Objectives, Discovery, Invention, Greenhouse, and Implementation/ Launch. Improve your chances of success and speed up your progress as you move through them by devoting far more time to the first three phases than you have in the past. If you consistently encounter problems in the latter two phases, consider whether you've tried to short-change the work of the first three phases.

- Increase your company's ability to innovate by using your well-managed innovation process to build people's trust in their own educated guts.

Note

1. James C. Collins and Jerry I. Porras, *Built to Last*, HarperBusiness, New York, 1997, pg. 89.

CHAPTER FIVE

Forming Your Innovation Team for Success

In business, creativity tends to begin with individuals, while the implementation of real innovation requires a team. Innovation needs input, collaboration, and lots of ongoing creative problem solving by a relatively large, cross-functional team of people. It also must have empowered leadership and decision making of an unusual kind. I will discuss real team empowerment in Chapter 6. This chapter explores team formation and decision making.

Forming your innovation team and determining how the people on it will make decisions are critical first steps in any innovation effort. It is at this point, however, that two new innovation roadblocks often arise. They are:

1. *People Myopia.* This occurs when management too narrowly defines the individuals who are chosen to participate on an innovation team.
2. *The Decision-Making Pendulum.* This is the struggle to balance autocracy and consensus on the team.

Now, before you skip to the next chapter because you believe you know all about the formation of cross-functional teams, think again. My experience is that most organizations are far too narrow and timid in their experience with this topic despite the fact that stretching the definition of what constitutes a great innovation team pays big dividends. I have never worked with an innovation team that included only the people initially envisioned by a client. Clients start out confidently telling us they have assembled a real cross-functional team, but we invariably need to prompt them to expand beyond their original view of who should be in the group.

Your Ideal Companions for Innovation

Ideally, your innovation team should contain a mix of three types of people:

1. *Champions.* Champions are the leaders of an innovation team. Good champions are visionaries who have a gift for recruiting those around them to work for and support their belief in something new. True empowerment and the ability to make motivating decisions are key to this role.
2. *Creators.* Creators germinate the newness. They see what Richard Farson calls the invisible obvious and help inspire the team with new, beginning possibilities. They also play a critical role in coming up with creative solutions to the hurdles that invariably arise during implementation.
3. *Doers.* Doers make new ideas real. Their creativity kicks in when it's time to put into operation the dreams of the visionaries and the creatives. Without them you are left with only ideas.

Your innovation team should begin with a core of people from each of these three categories. They will travel together through most of the journey toward newness. Also, keep in mind that during the development cycle an innovation effort must touch every function in your organization that is associated with the production, distribution, and marketing of your products or services. At certain points in your journey to innovation, you will need to solicit active participation from a resource team of individuals in every one of those functions. In this way, the makeup of your broader innovation group will vary over time.

Avoiding the People Myopia Roadblock

An effective innovation team marries expertise with both naïveté and diversity. It includes a critical mix of representatives from key stakeholder functions and seemingly irrelevant outsiders. By linking expertise with naïveté and diversity, this team has the resources needed to address any new idea, yet is not bound by the strict conventions of "how we do things around here." President John F. Kennedy once said, "Beware of experts!" He was right and he was wrong because, while they can unwittingly squash new thinking, experts are often critical to the implementation of new ideas. The trick is to not overload your team with them.

That's where the naïveté part comes in. Have you ever noticed how many scientific breakthroughs come from people who cross over from their own areas of expertise—the chemist who provides a breakthrough in physics or the botanist who identifies something new in aerodynamics? These people are modeling a vital element on any innovation team: those who come from seemingly unrelated and irrelevant worlds. At my company, we call them Wild Cards, and there is no such thing as an irrelevant Wild Card. In a few pages, I will give you some examples of how these people can help your innovation initiative. But first a few more thoughts about expertise.

Several years ago, I saw a wonderful ad in a magazine. The left page showed six experts who had made some very stupid pronouncements. Some of them were hysterically funny:

- "Heavier-than-air flying machines are impossible." Lord Kelvin, President, Royal Society c. 1885
- "Everything that can be invented has been invented." Charles H. Duell, Director of the U.S. Patent Office, 1899
- "[Babe] Ruth made a big mistake when he gave up pitching." Tris Speaker, 1921

Even that most revered of 20th-century experts and thinkers, Albert Einstein, was capable of monumental error, as Robert Youngson noted in his 1998 book entitled *Scientific Blunders*. Youngson quotes Einstein as having said, "There is not the slightest indication that energy will ever be obtainable from the atom."[1]

As these and numerous similar examples illustrate, relying only on experts can impede innovation because experts can sometimes suffer from myopia caused by the very expertise that makes them valuable to begin with. Often it is the nonexpert—the person without any preconceptions and prejudices—that produces the biggest breakthrough.

Peter Guber, producer of the movie *Gorillas in the Mist,* told *Fast Company* magazine in 1998 the story of how an intern saved his film with what at first seemed like a totally ridiculous idea.[2] His team was in an emergency meeting trying to come up with a solution for how to get the gorillas to act out the screenplay that had been written. They did not want to have to resort to the old, and largely unsuccessful, formula of using dwarfs in gorilla suits on a soundstage. Warner Brothers was financing the film and was worried that sending a crew into the Rwandan jungle to film over 200 animals would result in huge cost overruns. They were threatening to pull the plug on the whole movie project.

The problems seemed insurmountable until a young female intern asked a naïve question. "What if you let the gorillas write the story?" This query produced mainly laughter from the experts gathered in the room (humorous bazookas!).

Later, someone had the presence of mind to ask her what she meant. She replied, "What if you sent a really good cinematographer into the jungle with a ton of film to shoot the gorillas? Then you could write a story around what the gorillas did on film." Problem solved! They did exactly that and ended up shooting the film for half the original budget.

Such anecdotes, of course, do *not* mean that you should try to fill your team with only naïve people. Doing that can cause problems too, as one of my clients learned the hard way. Several years ago, this client approached all of the functional heads in the company, requesting the names of the most creative person in each department. The goal, which was a good one, was to form a cross-functional team that would be creative, break down walls, and make new things happen.

Unfortunately, what resulted was a wonderfully stimulating bunch of ideators. While generating lots of new thinking, they never got around to implementing anything. That's because the skills needed to break down walls and make new things happen go way beyond the ability to just think new thoughts.

The *M.A.S.H.* Analogy

One of the first things we do with clients is help them invite the right team of people to be part of their innovation effort. We like to encourage

them to envision the mix of characters who made up the medical crew of
M.A.S.H., the famous movie which then became a TV show that ran from
1972 to 1983. We want our innovation teams to function like that *M.A.S.H.*
unit, a wacky crew of disparate individuals who came together to do something important, each bringing a different talent to the unit when needed.

Like that *M.A.S.H.* gang, you definitely don't want an entire team of people who are only comfortable doing things "the way we've always done them"
and "by the book" (the Frank Burns and Margaret Houlihan types). But it is
wise to sprinkle a few of them into your cast of characters. They help keep
the team, and therefore the whole innovation project, rooted in a reality that
will be critical for effective implementation. They are often the critical doers.

The challenge is to not let the doers stifle the up-front speculation and
craziness of the creators, like Hawkeye Pierce and Max Klinger. In fact, we
often refer to those wild and crazy team members as our Klingers, and we
treasure them highly because they are hard to find and nurture in most
business cultures. (A corporate executive once told one of my colleagues,
"We kill the Klingers!") The Hawkeyes and Klingers don't panic—and may
even think it's fun—when the correct path to follow isn't totally clear. Such
people also are more fearless about breaking the rules; coloring outside
the lines is not a problem for them. You need people like this on any innovation team.

While it is important to take these paradoxical criteria into consideration in forming an innovation team (creators and doers; naïveté and expertise), it is also critical to eliminate the labeling of people as being either
one or the other. Ideally, if the team comes together and begins to work
effectively, you will find that some of the most creative ideas will come from
the doers (particularly in overcoming implementation roadblocks). Also, the
creators will be more willing to anticipate and deal with the implementation challenges earlier in the journey and less defensively. In effect, creators and doers will each assume the traits of the other at different times.
However, if you label people at the start, this transfer of abilities may be
lost because people tend to give you what they know you expect of them.

Don't Ban the Idea Killers

You should also consider another kind of organizational person for
your innovation team. I call them idea killers. They are famous throughout an organization for their long tradition of idea squashing.

I have observed an interesting phenomenon in my consulting career when it comes to these idea killers and the formation of an innovation team. Many times, at the start of a project, as we are discussing candidates for their innovation team, clients will assure me that "we really don't want Charlie or Betsy on the team." When I probe their reasons for saying this, the response is almost always rooted in a long history of this person blocking new ideas. Too often, when I explore that person's role in the organization, it is a critical one for effective innovation (usually the implementation part).

People like this should be included in your innovation initiative as part of the doers group no matter how difficult they are. Otherwise, they will sabotage the team from the bushes, and the key roles they play in the organization make it likely that they will prevail.

Here's another important lesson I've learned about idea killers. These obstructive people are usually quite smart, are very dedicated to their company, and are often as frustrated by the failure of past innovation efforts as everyone else. Their killer behaviors have arisen because little or no attention has been paid to group interaction, like establishing ground rules and other process techniques that facilitate open, honest communication and creative problem solving. The management of these behaviors is a critical success factor in all but the most innovative cultures, yet rarely, if ever, is the management of group interactions a defined part of someone's job on the team. The result is that idea killers are allowed to do their dirty work at will.

To rein in idea killers and turn them into true doers as well as to support the building of a strong and productive innovation team, you must use some form of group process facilitation. Of course, I believe in the power of professional facilitation. I've experienced it as a client and it's what I do now. But the designation, by any team, of at least one member (preferably a team of two) to focus his or her involvement only on processing the group's interactions is good. No matter how rudimentary the person's skills may be, the truth is that *any group process is better than no group process.*

Wild Cards Add Spice to Your Team

No innovation team is complete without Wild Cards. Wild Cards are just what they sound like. Some are experts, some are zany, unconventional thinkers from other business realms, and some are both! They represent noncompetitive, seemingly unrelated, yet analogous worlds. They bring fresh thinking and unusual perspectives to any group by simply sharing their experiences and stories during the critical exploratory, ideation, and implementation phases of any innovation effort.

Here are some examples:

- In a program on photo finishing, the client was encouraged to invite a dry cleaner into a technology exploration. Because photo finishing and dry cleaning share a common challenge in chemical waste disposal, the dry cleaner was able to bring all kinds of expertise and stories to the project.
- Over the years, I have become very fond of Dr. George Dorion, a brilliant, somewhat unconventional chemist. The good doctor retired a few years ago, but we like to keep him active from time to time as a Wild Card for other clients. A couple of years ago, we were working with a fabric company and urged them to invite Dr. Dorion to a technology exploration. They, of course, wondered how a chemist could help them develop fabric ideas. I responded that I had no idea. I just knew he was wonderfully stimulating, and I had a hunch he could bring something exciting and new to the project. They trusted me enough to invite him and fly him in for the session. I was not surprised (but they were) when he introduced them to the power of a technology that is frequently used in fermentation and showed them how it could be used in a breakthrough way in fabric. Eureka!
- In a project for Pitney Bowes, we invited former NASA physicist turned world-renowned portrait photographer, Mark Spencer, to an invention session on messaging simply because he represented world-class talent. His wacky, irreverent humor and unusual take on the world helped stimulate lots of new ideas.
- One of my oldest client relationships is with Bacardi. When they bought Dewar's Scotch, the management asked us to help develop a program to get 25- to 35-year-olds to "adopt" Dewar's. Because scotch is an acquired taste, like coffee, we arranged for Dave Olsen, one of our Starbucks Coffee Company clients, to participate in an early brainstorming session. Again, his analogies and stories were powerful stimulators of new thinking.

One final point about cross-functional teams. While you can find unlimited Wild Card resources outside your company, don't forget to look for the ones that are right inside your own company. A cohort of mine once worked with a petrochemical manufacturer who observed, "I've got people in the sky, people underground, and people in the deep, blue sea. And they never talk to each other!"

While everyone acknowledges the functional silos that crop up in every company, few ever address those silos that prevent cross-fertilization between seemingly unrelated divisions. Next time you are forming an innovation team think about the untapped resources in other parts of your corporate entity, the ones you are sure have nothing in common with what you do. Consider inviting some of them into your innovation effort. Many companies never think to call on their own diversity and, through this failure, miss out on some great ideas.

The Paradox of Championed Teamwork (Managing the Decision-Making Pendulum)

Now that you've thought about expanding your definition of the innovation team, let's talk about how decisions should be made as the team begins to work together. In my 30-year career, I have watched decision making swing like the pendulum in Figure 5.1.

When I took my first course in facilitation in the late 1970s, I was trained to serve the needs of the leader, the decision maker for the group. This focus on autocracy was great for facilitators because it provided clarity in decision making and speeded up the process. A good facilitator could oscillate between gathering open-minded input from the group and turning to one or two leaders for selection and decision making.

In theory, it worked very well. But, too often, after the rosy glow from the ideation session had passed, those same leaders were finding that their innovative decisions weren't getting implemented like they thought they would. Timetables weren't being met, the energy and support they thought they had created in the brainstorming sessions had evaporated, and team members who had seemed so supportive were now disgruntled.

This gap between decisions and implementation, which happens during any innovation initiative, has a number of causes. One of them can clearly be traced back to the team's decision-making practices.

Professional facilitators began to research the impact of autocratic decision making on group commitment and energy in the early 1980s. What we discovered was that the autocratic approach to decision making, while

[FIGURE 5.1] Decision-Making Pendulum

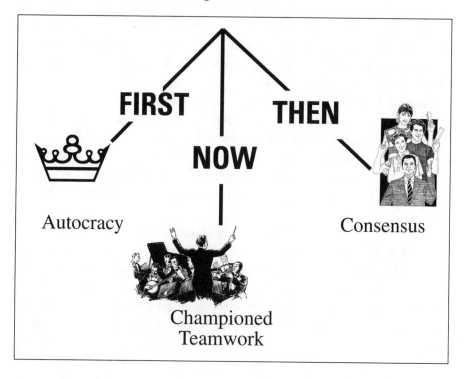

theoretically efficient, actually resulted in de-motivated team members who felt their creative ideas were being ignored and/or their issues were being discounted by the leaders.

Perhaps, because the timing of this awareness dovetailed with the growth of management practices like Total Quality Management (TQM) and Quality Circles, the decision-making pendulum swung all the way to the other side (as pendulums like to do), toward consensus. This was a disaster when it came to innovation for both the facilitators and the teams they worked with.

Consensus decision making in the pursuit of innovation tends to strip all of the creativity and excitement out of any potentially new concept. As explained earlier, there is no one way to achieve innovation. Therefore, every innovation team is constantly faced with many possible ways to proceed. If total consensus must rule in order to get everyone's agreement to move forward, any idea with real newness will automatically sink to its lowest possible common denominator by the time consensus is reached. In addition, the time to make a decision will become endless and the initiative will strangle on its own good intent.

If your mental pendulum were to settle midway between autocracy and consensus, it would stop at the point that I believe describes the best way to make decisions on an innovation team, a paradoxical technique we call Championed Teamwork.

Championed Teamwork describes the process by which decisions can be made most effectively by an innovation team. It requires both a high degree of trust between all of the members of the team plus total respect by the champions for the members of their team. It balances autocracy with consensus of the I-can-live-with-it kind, which is the only kind of consensus that will work in support of an innovation initiative. It also usually requires a process facilitator.

In a team that works under this model, the champions—ideally, no more than two people—make the ultimate decisions on major topics impacting the vision for the initiative. The hope is, of course, that everyone on the team will buy into and support the decisions of their champions. But, of course, it never consistently works that way throughout any change initiative. And, because there are few, if any, benchmarks by which to judge something truly new, decision making and teamwork can seriously break down when the team disagrees about what direction to take.

People in conflict tend to take a position and hold firm to it. Their mental viewfinders have a very narrow scope when it comes to problem solving. Everyone sees problems. Few are capable of open-mindedly discussing solution alternatives.

To address this natural tendency, the Championed Teamwork model sets up, in advance, for the champions to make informed decisions after problem solving with the team. It is here that the facilitator becomes so important. If there is strong dissension surrounding the resolution of a problem or challenge, the facilitator helps everyone explore the pros and cons of each opposing viewpoint. In this way, solutions can usually be found for the cons before the champions make their decision. If no resolution can be found, the team agrees to abide by and support the decision of the champions until some empirical evidence either supports the decision or suggests a reversal.

An effective Championed Teamwork model has many of the same strengths that make entrepreneurial organizations succeed. In companies that are still in the entrepreneurial stage, open discussion often thrives but there is usually one leader—the entrepreneurial founder—who will champion ideas and make the hard choices when they need to be made. Because the organization is new and "how things are done around here" has not been established, people are willing and able to thoroughly explore possibilities in a nondefensive manner. When companies that have moved far beyond the entrepreneurial stage want to pursue innovation, going back to this mode of operation with an intrapreneurial team that uses the championed teamwork model is extremely beneficial, but often difficult.

The true empowerment of the team and its leaders by senior management plays a huge role in the effective execution of the Championed Teamwork model. In Chapter 6, I will explain and show you how to resolve one of the biggest innovation roadblocks of all—the Leadership-Empowerment Fable.

[i n n o v a t i o n   f u e l]

- Understand that innovation depends on the interaction of people working in effective cross-functional teams. Expand your definition of cross-functionality to embrace the *M.A.S.H.* unit concept of a truly diverse group.

- Adopt a new decision-making model called Championed Teamwork, which is a facilitated balance between the two historic decision-making modes of autocracy (which frustrates teamwork) and consensus (which results in lowest-common-denominator ideas).

- Be sure your innovation team includes champions, creators, and doers. Know the strengths that each of these types of people brings to the effort so that you can design your process to make the most of each group's capabilities. But be careful not to brand people.

- Do not hesitate to add people to the team who have reputations as idea killers; managed properly, such individuals can be transformed into strong contributors.

- Do not succumb to People Myopia by relying too much on experts or by defining cross-functionality too narrowly. Reach out to Wild

Cards both within and outside your organization to bring as great a range of diversity of experience and viewpoints to idea development and problem solving as possible.

Notes

1. Robert M. Youngson, *Scientific Blunders,* Carroll & Graf Publishers, New York, 1998, pg. 13.
2. Peter Guber, *Fast Company,* June/July 1998, "My Greatest Lesson," pg. 83.

CHAPTER SIX

The Leadership-Empowerment Fable (Part 1)

Here's another all-too-common innovation scenario: It's early January. In a desire to promote fresh, out-of-the-box thinking, senior management establishes a task force to come up with some new, innovative ideas for growing the business in the next 12 months.

Hoping to encourage the team to start with a clean slate, senior management does not share any preconceived strategies or ideas with the team. Also, with the best of intentions, no one from senior management participates in any of the brainstorming to ensure that their presence doesn't disempower the team.

By late March, that well-meaning, highly motivated cross-functional team is ready to present an exciting array of challenging and innovative concepts to senior management. As a presumably empowered group, the innovation task force members assume that the purpose of this meeting is merely to update management on their progress to date.

Being professionals, the team prepares a thought-stimulating, buttoned-up presentation of its most provocative ideas as well as a plan to move ahead aggressively in order to stick to their very tight, 12-month timetable.

However, when the team's presentation ends, senior management begins to shred the ideas. They express concern over the absurdity of some of them, picking at every flaw. Surprisingly, they don't grasp the subtlety behind others. This response causes team members to become defensive

67

and somewhat hostile. Having assumed that they were truly empowered, they are unprepared for this reception.

At the same time, senior management wonders, "What have you guys been smoking?" and is appalled by the lack of strategic thinking that has resulted in a bunch of what they believe are hair-brained ideas. The situation deteriorates from there.

The Well-Intentioned Fable

This innovation roadblock is the Leadership-Empowerment Fable. The fable is a by-product of the creation of a false sense of empowerment. The sad part is that it results from nothing but good intentions. The fable comes into play when senior managers try to delegate to their innovation teams in the same way they empower their established businesses.

Between January and March, while that innovation task force was busy developing ideas, senior management was eagerly anticipating the results of its efforts. Unfortunately, both were unwittingly becoming caught between an ugly by-product of the Leadership-Empowerment Fable that I call the Corporate-Rain-Dance–Bat-and-Bow cycle.

Corporate rain dances are exercises in futility that waste energy without producing results. They occur when an allegedly empowered team engages in fun, stimulating strategy meetings and idea-generation sessions in which they often are extremely creative. But because of management's misguided attempt to make the group feel empowered without really giving them true power, the team invents in a vacuum, without any agreed-upon strategic direction and with an unrealistic view of the decision-making authority that has been granted to them.

The term *Bat-and-Bow* describes what happens when that allegedly empowered innovation team wraps up all of its inventive thinking in a slick presentational bow for its review with senior management. Usually, the more pleased the team is with the ideas (and with itself for having developed them), the fancier the bow becomes and the higher the team's expectations rise. The team hopes management will love everything and not change a thing. This is a surefire formula for disaster, and it can happen at any level of an organization.

Meanwhile, senior management, having honestly tried to empower team members by not influencing them with any preconceived ideas, too often comes to the review unconsciously expecting exciting, new, and highly feasible ideas that can be easily launched through established pipelines. Wanting to love everything, management understandably panics when, instead, the ideas presented are scary and have high degrees of teeth-gritting newness.

Unfortunately, the forum for this meeting only exacerbates the situation. The formality of the presentation and the love-it-or-hate-it mentality that governs idea evaluation in most organizations leaves management only one reactive option—to point out all the flaws in these concepts.

Metaphorically, management's reaction is like taking out a baseball bat and beating the crap out of everything—including the spirits, enthusiasm, and egos of the team members it wanted to empower. While this is most often unintentional, it is incredibly demoralizing for everyone.

In such situations, blamestorming replaces the brainstorming that should be occurring. The likelihood of disappointment, frustration, and no small amount of finger pointing on both sides of the table is almost 100 percent. Without ever articulating its assumptions, its objectives, or its own beginning ideas, senior management still expects the team to somehow understand what management wanted. Without realizing it, everyone has played a well-intentioned but very destructive game called "Guess what's in my head." Often this pattern repeats itself over and over, and, after a couple of Corporate-Rain-Dance–Bat-and-Bow cycles, good people stop wanting to participate in the Leadership-Empowerment Fable.

Preventing the Leadership-Empowerment Fable

There are really two mistakes that are routinely made in most companies, that contribute to the Leadership-Empowerment Fable:

1. The wrong decision-making structure for innovation
2. The total absence of an agreed-upon strategic vision for the business and an agreed-upon task focus for the innovation effort

In this chapter, I will discuss how to establish a flexible, workable decision-making model for the effective interaction of an innovation team with its senior management (at any level of the company). In Chapter 7, we will explore together how to set a motivating, empowering strategic focus for any innovation team.

How Decisions Get Made

To begin to understand how to overcome the Leadership-Empowerment Fable roadblock, you must first look at how decisions are made in business. Business decisions occur on three levels:

1. *Decision making.* People at this level have the power to say yes or no on major initiatives, especially for dollars and resources—either human or operational. Often one person but rarely more than three people.
2. *Decision recommending.* These are the advocates, often the team champions; often the implementers, too. Increasingly a miniteam of two to four people.
3. *Decision implementing.* The individuals at this level are the get-it-done people.

Every organization needs to have two models to empower and manage these three levels of decision making: one for handling decisions on established businesses and one for innovation-related decisions. Unfortunately, in most organizations only one decision-making model exists, and it is always the model for managing decisions on existing business. This is understandable because all the systems have been built to perpetuate and reinforce what already exists.

Early in my consulting career, I facilitated a number of Corporate-Rain-Dance–Bat-and-Bow cycles. This caused me to analyze the destructive forces at work to develop new ways to facilitate my clients' decision-making processes more productively.

What became immediately apparent was that the process for decision making on new ideas was all screwed up. Despite all protestations to the contrary, the decisions on innovative, new ideas were always being made at the top. And, in most of these companies, those same new ideas were having to first pass through several levels of decision recommenders who invariably were adding their modifications before the ideas finally reached the decision makers. By then, if even one thread of excitement from the original ideas still existed, it was a miracle.

My first inclination was to insist that senior management decision makers participate in the brainstorming and idea-development work. This failed for several reasons:

- First, despite the best of intentions, it is impossible for a cross-functional, multilevel room full of people to ignore the power bases in the room. I learned this shortly after I convinced my first senior manager to participate in a brainstorming session. At the start of the session, I cavalierly introduced the ground rule that declared "leave rank at the door." During the morning break, one of the participants informed me that she could leave rank at the door, but not her memory. She meant she couldn't wipe from her mind that her boss was sitting in the room and judging her ideas. That was my first clue that my solution might not be working.

- The second reason became apparent when I realized that the senior manager was, too often, less able to play because he or she was too far removed from the consumer and too mired in the status quo. This is not surprising. Did you ever notice how creative you can be in coming up with breakthrough solutions for someone else's problem? Have you also noticed how hard it is to deal open-mindedly with those same kinds of novel ideas when they are offered to address your task? This is at the root of the difference between decision recommending and decision making.

- The third clue showed itself when most senior managers simply refused to participate, allegedly due to a lack of time and too many other pressing priorities. My experience is that this is mostly true and partly Mahr's Law in operation (see explanation below).

My next strategy drove (I thought) to the heart of the decision-making process. If senior management was not going to participate, then the innovation team and its leadership had to be empowered to make decisions, even those that might appear foolish, wasteful, or absurd.

Guess what happened? Management promised to do this but, when the rubber met the road at decision-making time, it always usurped the decision-making power. Those reviews, as described above, automatically became decision-making meetings and, unfortunately, not of the collaborative, move-forward kind.

Mahr's Law

Corporate-Rain-Dance–Bat-and-Bow cycles are often the by-product of a history of failed innovation initiatives. Failure of any kind is a career-killer in too many companies. Regrettably, people who experience failure in corporate life are too often shifted over to assume strategic or new business development roles so that their careers can end respectably. This sends a loud, clear signal to any organization and results in what Synectics, Inc. cofounder, George Prince, humorously refers to as Mahr's Law of Limited Involvement. This law, which guides much of corporate behavior, has one basic rule: Don't get any on you!

In a culture that applies Mahr's Law to its innovation efforts, punishment can be meted out to anyone who comes in contact with any aspect of the innovation effort, *especially the decision making*. Is it any wonder, then, that the decision making surrounding innovative ideas would become so convoluted in most companies? I don't think so. Subconsciously, no one wants to touch it or be touched by it. That's why decision makers often hide from active participation in innovation (as hinted above), which fulfills Mahr's Law.

For me, helping companies identify their real decision-making protocols in support of innovation has become a major topic in my role as an innovation consultant and advisor because it is where the seeds of failure are so often sown.

One of the positive impacts of reengineering has been the removal of unnecessary layers of bureaucracy in many corporate structures. This has resulted in decision making being pushed down several levels below senior management. As expressed a number of times already, the infrastructure and systems that are created to support an existing business allow decisions to happen in this way, to provide checks and balances for those pushed-down decisions.

This can be a powerful, market-centered model for promoting better, quicker decision making on established businesses. It should work for most companies' lines of established business. However, it tends to create an unintended nightmare for those wanting to pursue innovation in those same companies.

Nothing is wrong with this decision model if the decision making, decision recommending, and decision implementing are made visible and managed properly. In fact, it would be impossible to operate any business without these levels of decisions.

But for innovation, decisions need to be managed more visibly because innovation does not have the same infrastructure supporting it as the day-to-day business does. Failing to establish clearly identified and involved decision-making, recommending, and implementing capacity, *at every level of the organization,* dooms business innovation. Everything hinges on this delicate balance between the innovation sponsors and team leaders who are truly empowered, capable of pursuing newness.

What Does Real Decision Making Mean?

When a client tells me the innovation team is truly empowered, I ask two simple questions:

1. If they invent something that is truly new (breakthrough or transformational), can this team spend large amounts of money and resources, *without your permission,* to develop it for market?
2. If an idea appears to be really big after development, can the team authorize capital expenditures and marketing budgets to support its launch *without your permission?*

If your answer to both of these questions (and it must be both) is yes, you may skip to the next chapter. If not, please read on. You can guess what my clients' answers have been for over 18 years. They have always been no. Why, then, call it an empowered team?

When decision making is positioned differently from its reality, as an innovation project or task force begins, you're instantly in trouble. As explained earlier, most senior managers truly want to empower their innovation teams. One of the not-so-hidden reasons for this is that, when it comes to innovation, senior management is often as unsure of where to find it as everyone else. So, they empower in the hope that someone at a lower level will discover Aladdin's magic lamp. Unfortunately, because of the way decisions are made on most innovation initiatives, management may never know that the magic lamp was ever in the building!

Structuring Decision Making for Breakthrough Innovation

In Chapter 1, I introduced some myths about innovation. Another myth is that a camel or a giraffe is a thoroughbred horse built by a committee. When a team creates a camel, it is not a reflection on the concept of teamwork but an indication that the team lacks critical elements that make teamwork successful. Unfortunately, at a time when organizations are increasingly being structured to benefit from the power of cross-functional teams, not only for innovation but in all aspects of business, too few people understand how to turn a group of camel designers into a group of thoroughbred horse creators. So the new emphasis on teamwork often produces less-than-anticipated results.

Two mistakes can cause this suboptimization of teamwork to happen in innovation work. One is that the team usually lacks a process leader to facilitate effective interaction. The facilitator helps make unbiased order out of the chaos that is always a critical element in every phase of innovation.

Mistake number two is the way decision making is inappropriately defined and mishandled. Fortunately, the answer to this decision-making challenge is deceptively simple; so simple that I fear its power might be missed. The only really tough part is procedural not conceptual.

The three secrets of an effective decision-making structure on any innovation initiative are:

1. Allow for no more than one reporting level to exist between the decision maker(s) and the decision recommender(s).
2. Agree on a lead facilitator who is trusted by both the decision maker and the team.
3. Make the decision implementer(s) part of the invention process.

The consequences of these three recommendations can be profound. If your company is honest about matching its objective for innovation with the proper level of decision making, it means that senior management will have to be involved and accountable for higher-level, breakthrough-related decisions.

If senior management makes this commitment to involvement and to decision making, it will have a demonstrable impact on any innovation effort. For one thing, it will create a highly visible model for the acceptability of well-intended failure at all levels. If an organization recognizes that disappointment is a frequent by-product of an innovative culture, this will significantly raise the bar on risk-taking permission. If people observe their management taking risks and experiencing acceptable consequences from those risks, they will be far more likely to try it themselves.

I once dated a young woman whose mother chain-smoked, yet threatened dire consequences if she ever caught her daughter with a cigarette in her mouth. "Do what I say, not what I do" was her mother's credo. It didn't work for her mother, and it won't work for you. If you want your people to take risks and pursue anything more than incremental innovation, you must be involved in it up to your eyeballs.

Why These Guidelines Work

Let's look at each of the guidelines I am suggesting.

Allow for No More than One Reporting Level to Exist between the Decision Maker and the Decision Recommender

As a manager, the first thing you must accept, if you are to properly structure decision making for breakthrough innovation, is that a labyrinthine decision-making structure (anything more than one level removed) will kill any prayer of achieving newness on your innovation initiatives.

Here may be one of Farson's invisible obvious observations: Because innovative ideas cannot be quantified in their formative stages (as previously argued), any modifications to an idea as it moves up the line is just that— a modification. True innovation comes from passionate champions, not consensus. Any additional links in the decision-making chain just add layers of modification to any recommendation. Too many voices modifying a beginning idea will destroy it. This will be true no matter where the decision making resides, whether at a senior management level or at an operating, brand, divisional, or business unit level.

The corollary to these two rules is that the higher the goal level you set for innovation, the higher the level of senior management that will need to be involved, yet still no more than one level removed.

To do this, you must be truly honest in determining the level of innovation to be pursued—incremental, breakthrough, or transformational. If your goal is incremental innovation (now be honest!), then you can probably look to the lower levels of your organization to manage it, but *only* if that group never has to come to you for approval. Remember that the likelihood of failure for any type of innovation initiative will significantly increase if more than one reporting level exists between the decision recommender and the decision maker.

This approach can also significantly accelerate the innovation process in large organizations by eliminating those unneeded, destructive levels of modification through which ideas usually must pass.

The test for this is easy, and it goes back to my two earlier questions:

1. If they invent something that is truly new (breakthrough or transformational), can this team spend large amounts of money and resources, *without your permission*, to develop it for market?
2. If an idea appears to be really big after development, can the team authorize capital expenditures and marketing budgets to support its launch *without your permission?*

You will know you've reached the right level of decision making on any innovation initiative when the answer to both of these questions is yes.

Please understand that I am not advocating for management to be part of every phase of any innovation program. I'm also not suggesting that the whole innovation team should be comprised of people only one level down from the decision maker.

I always push for the inclusion of people from every level and discipline in the company. I'm referring to the link between the decision recommenders (who become the leaders of the innovation team) and the decision maker. That link can only be one reporting layer removed.

As to the involvement of the decision maker, I actually prefer to keep management out of most of the work because of the tendency to stifle people in the strategy and brainstorming sessions and the tendency to be farthest removed from the consumer.

Agree on a Lead Facilitator Who Is Trusted by Both the Decision Maker and the Team

This is where the facilitator becomes important. As mentioned in Chapter 5, I strongly advocate the designation of at least one (preferably two) of the members of the team to be process facilitators, totally absolved from any content-related responsibilities. These might be professional facilitators (which I would, of course, recommend), a team from your human resources and/or training department (if they are trained in these skills), or just people-oriented members of the group.

The loftier the goal, the more important the skills needed to help you and your innovation team get where you want to go. Well-trained facilitators will help the initiative get off on the right foot. As the project begins, they will help both parties to:

- Agree on the focus and parameters of the project
- Determine the timetable and measurables
- Clarify roles and ground rules for effective interaction, especially for those places in the process where the decision maker needs to have a voice

As the project ramps up, the facilitators will know when it is appropriate to reconnect the team with the decision maker. This will be especially important at those critical junctures in any innovation initiative where options are being selected for further development. These might be strategic options, idea development options, or testing options, among others.

If the facilitators have the trust of both the team and the decision maker, they will be able to initiate what I like to call shirt-sleeve sessions, as opposed to Bat-and-Bow reviews. Shirt-sleeve sessions are informal working sessions that are critical for breaking down the love-it-or-hate-it mentality that always accompanies formal reviews.

In *Built to Last,* Collins and Porras describe the difference between the tyranny of the *or* versus the genius of the *and.*[1] It took me a while to grasp their point until I connected it to something I've been teaching for over 18 years. Truly innovative ideas are almost always born with great promise and serious flaws. If you can only love *or* hate a fragile, new idea, you will most likely choose to hate it. But what if you could both love *and* hate a new idea? Might you not be more open-minded to some creative problem-solving techniques that could help keep it alive?

Helping people adopt the *and* and drop the *or* approach to idea evaluation is one of the most important roles any facilitator can assume on an innovation team. For example, let's return to the scenario at the beginning of this chapter. Imagine if the Bat-and-Bow review hadn't happened as I described it. What if a facilitator had kicked off the meeting with a few ground rules to set the stage for informal interaction versus rigid presentation? What if the team had been set up, way back when the project began, to expect and value input from the decision makers? What if the decision makers had been facilitated through an open-minded evaluation of the ideas being presented, able to both love and hate the same idea, in a way that invited problem solving instead of idea bashing? (I will introduce skills to accomplish this in Chapter 13.)

These are the talents that a worthy facilitator will bring to an innovation team. And the paradox of good facilitation is: The easier it looks, the more skill it takes.

Make the Decision Implementer Part of the Invention Process

Here's another all-too-frequent scenario: One work team creates some exciting new ideas for its product, service, brand, or business. It is then required to pass those ideas off to a separate group for implementation.

I can't count the number of potentially breakthrough ideas that have fallen through the cracks at this critical point in the developmental process, because of this hand-off to another team. These ideas invariably experience the not-invented-here syndrome and die aborning.

This is often the way organizations have been structured—to separate the strategists and ideators from the implementers. Years ago, it was less of a concern because most of my clients looked to their invention teams to implement their own ideas. But the hand-off approach is far more of a concern today due mainly to the increase in matrix work structures in almost every industry including, most particularly, the high-tech field.

The solution to this hand-off, not-invented-here dilemma is to involve members of the implementation team in the strategy-ideation process. While this may appear to be a no-brainer, you'd be amazed at how few do it. It not only fosters teamwork and an enthusiasm for having participated in what's coming, it also heightens the likelihood that major implementation issues will be dealt with during the strategy and ideation phases. Remember, like TV's *M.A.S.H.* unit, your team needs a blend of champions, creators, and doers.

As we leave this topic, I just have to reiterate the destructive power of the Corporate-Rain-Dance–Bat-and-Bow cycle that is at the heart of the Leadership-Empowerment Fable. True innovation empowerment should be easier to achieve. The challenge is to match the level of the innovation goal with the appropriate level of decision making. There's no faking it, yet I don't see too many organizations really doing it properly.

Decision making structured for real innovation is the first key to truly empowering an innovation team. The second key is to provide the innovation team with clear and motivating direction and objectives. I'll discuss the important role that strategic visioning and goal setting plays in eradicating the Leadership-Empowerment Fable in Chapter 7.

[i n n o v a t i o n   f u e l]

- Realize that the decision-making process needed to support innovation success differs fundamentally from the decision-making process used to manage your existing business.

- Put in place an innovation decision-making structure with no more than one reporting level between the decision maker(s) and the decision recommender(s). Have the decision recommenders lead the innovation effort.

- Avoid Corporate-Rain-Dance–Bat-and-Bow cycles, in which allegedly empowered teams develop ideas only to have them crushed when they don't meet management's expectations. Eliminate this problem by providing a clear, visible decision-making process and a strategic vision and innovation task focus from the start.

- Use skilled process facilitators who can develop a strong level of trust with both the innovation team and the decision makers.

- Make sure that the people who will have to implement the ideas that come out of the Invention phase of innovation are part of your team throughout the entire process. This will help eliminate the not-invented-here thinking that often impedes the implementation of breakthrough new ideas.

Note

1. James C. Collins and Jerry I. Porras, *Built to Last*, HarperBusiness, New York, 1997, pg. 44.

[PHASE ONE]

Setting Objectives

CHAPTER SEVEN

The Leadership-Empowerment Fable (Part 2)

Many truly innovative ideas crash and burn on the road to implementation because, at the start, organizations fail to set meaningful objectives by which to evaluate them. The result has been described as "Ready-Fire-Aim." Unfortunately, too often, the significance of setting objectives for any innovation initiative does not become obvious until the project is way down the road. Only when it is time to deal with the selection and implementation of new ideas does it become apparent that you may not have set a clear direction at the beginning of the effort. Trying to skip this vital step will inevitably lead to big problems at every subsequent stop along your innovation journey.

Ready-Fire-Aim is an innovation roadblock that arises out of the same misguided management behavior that creates the Leadership-Empowerment Fable—the reluctance to tell the innovation team what management really wants for fear that this will inhibit ideation. Holding onto this faulty conviction—that unrestrained and unfocused innovation will have the best chance of achieving real breakthroughs—is a serious mistake. The truth is that business innovation succeeds best when it has a clearly articulated and motivating focus.

Some companies resist setting objectives for innovation because they believe it is a waste of time. They are sure their innovation destination will somehow magically appear along the way, making it unnecessary to decide in advance where they want to go. This type of thinking tends to happen to people who adhere to one of the myths dispelled in Chapter 1, the mistaken belief that breakthrough ideas will be obvious when they first appear. Buying into that myth makes you less likely to understand the need to set goals for your innovation effort before starting.

Why Objective Setting Is So Critical for Effective Innovation

The problem with not setting objectives is that, to paraphrase the Cheshire Cat in *Alice in Wonderland,* "If you don't know where you're going, any road will take you there." Another old adage says, "Aim at nothing and you will succeed."

The only way to avoid an ultimately frustrating Ready-Fire-Aim innovation effort is to have (1) a motivating, empowering strategic vision for the business, and (2) a challenging, focused, and exciting task focus for any innovation effort meant to support that vision.

These two elements are defined differently, yet complementarily, to work in concert with each other. If defined and agreed on by all parties, *before* the innovation effort begins, these goals and measurable milestones can drive the periodic checkpoints that should happen all along the innovation highway. By establishing these fundamental guideposts at the start, you avoid the fractious and time-wasting debate that can arise at numerous points down the road if people aren't on the same page right from the start.

In other words, a strategic vision for the business and a task focus for the innovation effort don't just start the innovation team off on the right foot; they also serve as touchstones that protect the entire process along each subsequent step of the innovation effort. These touchstones enable management to truly empower innovation teams in ways that virtually wipe out the guess-what's-in-my-head? game. With these powerful, clear-headed, motivating objectives as decision guides for both the team and its managers, it's easier to overcome roadblocks all along the innovation highway.

Setting Objectives is important for any innovation program because there are few tools in existing business systems with which to judge new ideas and make decisions when trying to effect change. Some executives think a wishy-washy statement like "think out of the box" or a financial measurement like "X dollars of revenue from innovative new ideas within three years" should provide sufficient guidance and motivation for the innovation team. Unfortunately, in most organizations, it doesn't.

Other management teams go through the motion of setting innovation goals but fail to ensure that all of the elements that will support the achievement of those objectives are in place.

Think back to the scenario I introduced in Chapter 6 where the innovation team and management went through a Corporate-Rain-Dance–Bat-and-Bow cycle. The mistakes described in that frustrating reality are made, so often because management doesn't understand the importance of the Setting Objectives phase of innovation or how to go about it. Management doesn't know the difference between telling the team which direction it wants the team to go (good) and telling the team what it wants the team to do (bad). The former is inspiring and very motivating, while the latter thoroughly disempowers and demotivates people.

Phase 1's Critical Success Factors

During my career in innovation, I have identified factors that are critical to the success of each phase in the innovation process. I will share these with you as we explore the five phases of innovation throughout the remainder of this book. Here are the critical success factors that produce and support effective objective setting for innovation (Phase 1):

- A clearly-articulated, shared, and motivating strategic vision (core ideology and envisioned future) for the business (whether that be the entire corporation, a business unit, a brand, or any other subunit)
- A clearly defined task focus for the innovation effort
- Predetermined measures for success
- Clearly identified and involved (where appropriate) decision makers, decision recommenders, and decision implementers
- A multilevel, cross-functional core team involved from start to finish

The last two critical success factors (decision-making structure and cross-functional team) were discussed in Chapters 5 and 6; the rest of this chapter covers the first three (strategic vision, task focus, and success measures). All are critical to successfully avoiding the Ready-Fire-Aim roadblock, which you are likely to meet down the road if Phase 1, Setting Objectives, is not completed properly.

The Vision Thing

Vision has become one of the most abused terms in business. Too often, it is confused and even used interchangeably with the term *mission*. Whole books have been written about this topic, but most companies still don't get it.

Misperceptions about what makes a good vision show up in different ways. Often, when I ask if new clients have a strategic vision for their business, they say, "Oh, yeah, we've got one." They point to a plaque on the wall that is six paragraphs long. The information on this plaque too often is a mishmash of their vision, their mission, and their core values. Other companies think a vision is just a slogan developed for consumer consumption.

In *Built to Last,* Collins and Porras provide the best rationale and definition for a strategic business vision that I've ever encountered, and with Jim Collins's blessing, I will draw liberally on their viewpoints here. As they state, an inspiring strategic vision is comprised of two pairs of somewhat paradoxical elements: a core ideology and an envisioned future.[1]

Elements of a Strategic Vision

A. Core Ideology
   1. Core Purpose
   2. Core Values
B. Envisioned Future
   1. Big, Hairy, Audacious Goal (BHAG, pronounced bee-hag)
   2. Vivid Description of the Future

If properly crafted and aligned, these elements of a strategic vision can drive innovation in ways that nothing else can. And here's another important point: If you accept the proposition that the established business systems for judging new ideas don't work, then you should appreciate how critical these elements of the strategic vision will become as your guideposts for innovation.

While I will elaborate on these components of a strategic vision throughout this chapter, I strongly encourage you to read *Built to Last* for the definitive explanation of these critical elements for supporting and guiding breakthrough innovation. Since devouring that book, I now ask my clients' innovation teams to read it before we embark on their Phase 1 work. In every case, this has made our work in this critical phase more productive and easier because people come to the table with a shared understanding of the critical elements of a strategic vision and how these elements can drive innovation. I can guarantee that having this shared knowledge will prove valuable to your innovation team, too.

Interestingly, not having a well-stated strategic vision for a business doesn't always become critical until you start to innovate for that business. Many organizations move along quite smoothly for long periods of time without one. That's because every corporate culture communicates an unstated strategic vision in a million different ways, which reinforce how the company does business. In other words, operating without a visible strategic vision reinforces the status quo.

How Encompassing Must Your Business's Vision Be?

In concept, it is possible to create a strategic vision (core ideology and envisioned future) for an entire business entity or for divisions of that entity right down to a unit or brand level.

Quite simply, the scope of the strategic vision should match the goal level of the innovation initiative. By that, I mean that a breakthrough innovation effort for a whole company can only be guided by a strategic vision for that entire company. On the other hand, a breakthrough innovation effort for a business unit or brand can be guided by a strategic vision for either that business unit or brand.

Core Values

The starting point for developing your vision is to define your core values. Core values, the first elements of core ideology, are the touchstones that help the innovation team make decisions. Again, I refer you to *Built to Last* for the definitive explanation of this key element of any strategic business vision. There is nothing I could offer that would be anything more than pale redundancy. But please don't take this paucity of words on my part as an indication that core values are something you can skip. They are a vital part of any successful innovation effort.

Getting the Core Purpose Part of Vision Right

Let's look at some great core purpose statements. The ones listed below belong to four of the visionary organizations reviewed in *Built to Last:*[2]

- "To strengthen the social fabric by continually democratizing home ownership"—Fannie Mae
- "To preserve and improve human life"—Merck
- "To make people happy"—Walt Disney
- "To give ordinary folk the chance to buy the same things as rich people"—Wal-Mart

These highly effective core purpose statements have several characteristics in common:

- *Great core purpose statements help define what business a company/organization sees itself as being in (and, by inference, not in).* A great core purpose (sometimes referred to as a vision statement, although I believe Collins and Porras might disagree) defines an organization's fundamental reason for existing. A core purpose is *not* meant to be a statement of differentiation (i.e., K-Mart might believe it has the same core purpose as Wal-Mart). A core purpose also helps an organization identify what it is *not* about, which becomes very important when evaluating potential ideas for innovation.
- *A great core purpose is championed by senior management.* A strong core purpose signals the troops that the business leaders have agreed on a clear course for the "corporate ship," to both maintain the core business and pursue breakthrough newness.
- *A great core purpose has no time frame or other numerical limits.* It puts forth an unreachable and inspiring aspiration. Unfortunately, these non-numerical attributes of a core purpose, which help a strategic vision to be effective in terms of motivating people to drive toward the future, also make some management teams very uncomfortable. These executives avoid the whole subject of core ideology and envisioned future because they think the concept is too touchy-feely. They can't relate to the notion of objectives that don't have a time limit, a desired monetary outcome, or other quantifiable elements attached to them.
- *A great core purpose is not meant to be seen by consumers.* None of these visionary companies ever developed their core purposes for consumer

consumption. Nevertheless, each articulated a very effective core purpose in establishing its strategic approach to the marketplace. And, while there would certainly be nothing wrong if consumers did see them, the point is that they weren't developed with a mind-set for them to become slogans.

In the words of Neal Thornberry, Professor of Management at Babson College, "Vision statements should be designed to be vivid, memorable, inspiring, meaningful, and brief. . . . It is the part that gives direction, helps focus effort, and stays etched in one's mind."[3]

For a core purpose to work, people have to be clear about exactly what it means. This sounds simplistic but can actually be hard to achieve. People can look at the same words and understand them quite differently. Just think how many times you have been involved in a maddening conversation where you and the other person were using the same words only to discover, after a lot of confusion, that you were defining those words very differently.

For example, my firm's core purpose, our fundamental reason for being, is: Facilitating breakthrough innovation that works. It took us almost two days to agree on these five words. The first three came very quickly. The real debate surrounded the last two. When we crafted this statement, one of our colleagues was a fellow named Jay Terwilliger. Jay is one of the most gifted, instinctive facilitators with whom I have ever worked. In his brief two years with us, he had an impact on virtually every aspect of our business, including our strategic core purpose.

Jay came out of the advertising business. He proposed adding "that works" to our core purpose based on his experience with allegedly breakthrough advertising that never worked. He felt that our commitment to the realities, the implementation aspect of innovation, was so strong that it needed to be stated in our core purpose, even at the risk of perceived redundancy.

I, for one, didn't get it at first. But Jay was so passionate on the topic that he won me, and eventually everyone else, over to his position. Now, after living with it for so many years, I can't imagine it not being a key element of our core purpose. I also believe that it's part of what makes us different.

There was also another challenge to the adoption of this core purpose for our company. In working it through, we came to realize that, for some of our staff, *facilitating* meant getting up in front of a room and guiding meetings. For me, it meant empowering people to achieve innovation through a variety of behavioral tools and techniques, both in and out of meetings. If we hadn't discussed what everyone in the room understood facilitating to mean, we would not have ended up with a shared core purpose that survives and motivates us as a strategic touchstone to this day.

A Classic Example of Why Words Matter

Let me give a simple yet powerful example of why it's important to be sure everyone has a shared business definition and is therefore headed toward the same goal line. Many years ago (before I had ever heard of Collins and Porras, core purposes, or *Built to Last*), I was interviewing a cross section of the management of a frozen fish company. Each interview was held one-on-one for about one hour.

During my hour with the division's president, I asked him, "What business is this company in?" His answer was immediate and firm. "We're in the food business." I was a bit surprised by the expansiveness of his response because, historically, this company had played only in the frozen segment of the food business.

His reaction to my surprise was unequivocal. "Our parent company has 50 operating units that cover every part of the plate." (*Part of the plate* is a food industry term to describe different competitive segments, such as appetizer, soup, salad, entree, starch, vegetable, and dessert.) "If we come up with something truly innovative, we could achieve distribution through any one of those subsidiaries."

I then interviewed the marketing vice president who informed me, in no uncertain terms, "We're in the frozen food business. Make no mistake about it." He indicated that he had absolutely no interest in considering ideas outside of this distribution frame.

The head of R&D told me, "We market frozen, breaded slivers of fish." And a product manager concluded, "We market food products meant to be dipped in ketchup."

With these four diverse statements, I was able to explain the importance of having a strategic business definition for guiding this business unit's innovation effort. If we were to begin without aligning these four views we would be doomed to failure, because:

- The president's business definition ("the food business") was too broad for his troops to contend with.

- The marketing guy's statement ("the frozen food business") was potentially too narrow, especially if the president was expecting ideas that went beyond current frozen food distribution channels.
- The head of R&D's view ("frozen, breaded slivers of fish") was so rooted in history and his technological comfort zone that it would have stifled almost anything new.
- The product manager's definition ("products for dipping in ketchup") was a totally different take from the other three and was, in its own way, also extremely narrow and limiting.

This is only one of many experiences that have demonstrated the importance to me of beginning any innovation initiative with a multilevel conversation about the strategic vision for the business.

Our Experience with Core Purpose Statements

At Creative Realities we have achieved quite a bit of success helping clients craft stimulating and motivating core purpose statements. Some of the ones we're proudest of (and that I've been given permission to share) include:

- "Fun/ctional, non-electric tools and gadgets for the kitchen person" —Ekco Housewares
  This is a creative play on words (fun and functional) that focuses, ultimately, on the end user's mind-set yet is defined in business terminology.
- "Up and down safely"—Werner Ladder
  This is short and sweet, and it prompts thinking that takes the company beyond just the ladder market.
- "To bring out the best in anyone . . . any time, anywhere"—eMaritz Inc.
  A world-class evolution from traditional, big-company performance improvement incentives to the promise of individuality as can only be inspired by the Internet.

Seeing the Future

You are now halfway toward completing your Setting Objectives phase. The first two elements of the core ideology (core purpose and core values) are meant to be the unchanging foundation upon which the organization is built. The remaining elements, which comprise the envisioned future, are meant to drive change. These include: (1) a big, hairy, audacious goal (BHAG, pronounced bee-hag), which is especially important for galvanizing an organization, cross-functionally, behind an innovation effort, and (2) a vivid description of the future, which will inspire people by painting a mental picture of what the world will look like when the BHAG is achieved.

The BHAG/Task Focus

Let's be clear. Shareholder wealth never motivated anyone who wasn't a shareholder. And innovation goals focused on return on investment (ROI) are equally demotivating to all but the most hardened financial soul.

That's why I got very excited when Collins and Porras introduced the concept of the BHAG.[4] A BHAG bites off part of the vision and attaches time and other measurability factors to it. A good BHAG is risky but attainable within a defined time frame. Management must stretch to set a BHAG, and the whole organization will need to pitch in to make it happen. You must be somewhat arrogant to tackle a BHAG, convinced that you can do it. And, like the other elements of a strategic vision, a BHAG can be used to challenge an entire organization, division, business unit, or brand. But unlike the other three elements of your strategic vision (core purpose, core values, and vivid description of the future), which are inspiringly unmeasurable, a BHAG has a deadline (usually around ten years) and performance measurements.

President Kennedy's challenge that the United States would land a man on the moon and return him safely to earth by the end of the decade was a BHAG with which most people are familiar. (Unfortunately, NASA has floundered since that BHAG was achieved, largely because NASA didn't have a core purpose driving that BHAG. I've always thought, for instance, that Gene Rodenberry's core purpose for *Star Trek*—"to go where no human has gone before"—would make a great core purpose for NASA.)

Some other famous BHAGs are:

- "Bet the pot on the 747"—Boeing (1960s)
- "Democratize the automobile"—Ford (1910)
- "Crush Adidas"—Nike (1960s)

- "Become the Harvard of the West"—Stanford (1940s)
- "Become a $125 billion company by the year 2000"—Wal-Mart (1990)[5]

A lesser known, yet powerfully effective BHAG for the former Massachusetts bank, BayBanks, was "to become the McDonald's of banking." With this business goal it captured a large share of the consumer banking market in my home state. The clear implication of a McDonald's business metaphor suggested low-cost consistency, access on almost every street corner, and ease of use. It also implied that you wouldn't be able to have it your way. And that's just the type of banking organization BayBanks became before they were gobbled up in a series of mergers.

The reason I love BHAGs is that, like core purpose statements, they are inspiring and motivating as opposed to most mission statements I've seen. Mission statements tend to revert back to market share and ROI goals, which are not tremendously motivating.

I also love the BHAG concept because your innovation task focus should be a short-term version of a BHAG. They could be one and the same. More and more companies are using them to galvanize their entire work forces in support of their loftiest innovation goals.

To complete the Setting Objectives phase of your innovation initiative, you will need to craft a task focus for your innovation effort. I like to call this "your one-sentence invitation to innovation." Like a BHAG, a well-crafted task focus sets a lofty yet achievable goal for any innovation effort. Many choose to identify a shorter-term innovation task focus, usually in support of a longer-term BHAG and in never-ending pursuit of fulfilling the core purpose.

The innovation task focus that evolves from a business's core purpose, core values, BHAG, and vivid description of the future is a goal that is achievable in a shorter time frame than most BHAGs, usually within two years. Achieving this task focus will make a significant difference in your ability to fulfill your vision.

Innovation teams run into problems defining an appropriate task focus for innovation in several ways. First, too many groups stop after defining their business core purpose (or vision statement). They think this should be sufficient to drive innovation. It's not. It's much too broad to give people the direction they need for moving forward in a unified way.

Other teams carve out a task focus that supports the vision but is either too broad or too narrow. Again, if the task focus is too broad, people will feel overwhelmed and not have enough guidance about how to approach the assignment. On the other hand, if you define the task focus too narrowly, it will limit creativity and decrease the odds that breakthrough ideas will be generated and pursued.

Because innovation task focuses are so proprietary, it is difficult to share any from my clients. However, here are some older examples that should give you an idea of what constitutes a well-defined task focus:

- "Develop a regional distribution system for delivering fresh-prepared [food products] to retail outlets daily."—A global food marketer
- "Create a plan to retain 80 percent of the revenue generated from our successful CD offering before the CD's term is up in six months." —A regional bank
- "Invent the next generation of commercial desktop computers."—A computer company

Vivid Description of the Future

A well-crafted vivid description of the future is, in essence, a one-paragraph scenario of the future of the organization. As I said earlier, it describes what the world will look like when the BHAG is achieved. It creates mental pictures with lofty, inspiring images about how the organization will interact with itself, its consumers, and its competitors in its marketplace environment. (Again, see *Built to Last.*)

A good vivid description shows people what can be, which is why it is critical for innovation. However, some companies develop the verbiage but run into problems because of constraints that preclude them from pursuing their envisioned future. These barriers can be either actively stated or can be implicit in the culture. Every organization has a set of unwritten rules, for instance, that encourage or inhibit certain behaviors. These unwritten rules are very powerful and will win out over the best strategic vision ever articulated. If this happens, then your strategic vision (core ideology and envisioned future) will be worthless.

Part of the challenge for any innovation initiative is to identify the cultural barriers that cause this misalignment and help bring them out into the open where they can be addressed. Only in this way can all of the ingredients that make up an inspiring strategic vision be aligned with a reality and culture that supports, instead of blocks, the direction in which you want your organization to travel.

[FIGURE 7.1] Creating Your Core Purpose and BHAG/Task Focus

Here is a synopsis of how to create a core purpose and a BHAG/task focus:

Core Purpose

(The organization's fundamental reason for being)

Innovation BHAG/Task Focus

(A two- to ten-year stretch, measurable bite of the core purpose)

Criteria for a Core Purpose

- Guiding philosophy of the business we're in (and not in)
- Sets sights high
- No finish line or time limit
- Brief and memorable
- Not a slogan

Criteria for a BHAG/Task Focus

- A challenging, measurable goal at which we can succeed
- Will have meaningful impact toward achieving the core purpose
- Requires cross-functional teamwork to achieve

Setting Success Measures in Advance

Once clients have crafted their core ideology and envisioned future with a challenging BHAG and innovation task focus, I ask how they will know they've succeeded with their innovation effort. A common response is, "Oh, we'll know it when we see it." Or they develop success measures that are so narrow they constrain innovation. Often this comes in the form of deciding that the outcome of any innovation effort must match the standards used for existing lines of business in terms of volume, profit, and other measurements. These opposite approaches—going with no measurements at all or setting measurements that are too restrictive—can cause a lot of innovation pain.

Pursuing the first route—not setting good measurable standards for success at the beginning of any innovation initiative—means people (and their organizations) won't be able to know if they have succeeded. For instance, let's say Donita is put in charge of managing an innovation effort; she is heading your team. If none of her team's aspirations are tied to any overall measurables for her business unit, then it is likely that Donita will not get much support when she tries to elicit developmental support from R&D, or when she goes to Marketing when she wants to visit the field to talk to consumers. If no one else is being held accountable for helping her (and thus supporting the innovation effort), the situation will become unworkable and no progress will be made.

Beware, however, of pushing the pendulum too far in the other direction. You don't want to set up situations in which the plant manager comes to Donita and suggests running an unnecessary test because he needs more innovation time on his record to receive his bonus. Instead, you need to attach realistic business goals to the effort that promote innovation through a broad sharing of responsibility, yet don't encourage abuses of self-interest.

Paradigm Confusion

Here's a story that illustrates the kind of problem you can get into if you take the second route I mentioned above and set innovation success measurements that mirror those of existing lines of business and are therefore too restrictive. Back in the late 1970s, I was recruited by a nationally recognized, over-the-counter (OTC) health care company that wanted to venture into the food business. The company was going to purchase a health food store line of products and, because of my new venture experience at Colgate-Palmolive, they wanted me to help launch this acquisition into supermarkets.

I quickly discovered that the existing business operated with two paradigms that totally conflicted with those of the food business they wanted me to create.

The first paradigm revolved around product taste. In their established lines of business (OTC medicines), the product was expected to taste bad in order to be good for you. Everyone is accustomed to medicine tasting bad. When I pointed out that this line of health foods also didn't taste very good, management's response was, "Americans are becoming more health conscious and they are going to be willing to sacrifice taste for health benefits." It's now over two decades later and that prediction still hasn't proven true. But the rationale "fit" their business mentality.

The second paradigm was around profit margins. The company was used to operating with a cost of goods that was no more than 25 percent. This allowed for at least 55 percent of the gross profit margin to be spent on advertising and marketing, which was a critical element in their OTC tool box.

In developing the business plan for this new food business, it quickly became apparent that the cost of goods was going to be at least 40 percent, which I soon discovered (after some benchmarking research) was common in the food business. This would mean that my OTC medications employer would need to accept a new volume-profit paradigm, spending much less for advertising and marketing (as was the norm in food) and living with a higher-volume/lower-profit-per-item business model than they were used to.

When I explained to the president of my division that the food industry operated with a different business model, his response was, "We're gonna teach them the drug business." He thought the company could change the food industry's whole business model! Needless to say, the food division never got off the ground because the company failed to recognize that, as management guru Peter Drucker says, "If you want to do something new, you have to stop doing something old."

Requiring that a breakthrough innovation effort fit within those business paradigms that you are comfortable with doesn't always work. It severely limits creativity from the get-go because, when such a mind-set is prevalent, people will avoid putting forth ideas that don't fit the mold. It also creates huge amounts of frustration and wastes resources because ideas that do move toward implementation will run into repeated roadblocks imposed by a management with a rigid mind-set about "how things should work around here."

In some cases, success measurements are simple to define. For instance, a merchandise group once asked us to work with them to create some differentiation in their theme park retail shops, which were becoming too homogenous. This meant people had little incentive to explore other parts of the park when they were buying their souvenirs. The clear measure of success for this effort was that the strategies that were developed started to drive people to shops in different parts of the park.

When to Apply Your Success Measures

Another way in which companies and their innovation teams get into trouble with success measures is by applying them too early in the innovation process. While you must define success measures at the outset, you don't introduce them into the process until near the end of Phase 3, Invention, when you are at the point of developing your beginning ideas into innovative concepts.

If you introduce success measures before people even begin brainstorming, you will restrict their thinking. But if you wait until after the ideas you're working on have developed some legs (momentum) and generated enthusiasm and interest, you can introduce those measurement criteria when they are appropriate.

We have a very revealing training exercise that reinforces this point. At the beginning of any client's training course in creative problem-solving skills, we run a teamwork exercise before we teach them anything. This involves a 30-minute meeting, which the group is asked to run on a problem needing creative solutions. We provide a problem that states a real-life situation and three to four criteria for success.

For example, my first business partner found an old magazine article that raised an issue about the plight of seals that were becoming entangled in fishing nets. This was resulting in many drowned seals and damaged fishing lines. While the article never stated any criteria, we tell our teams that their solutions:

- May not cost more than $100 per fishing boat
- Must be environmentally benign
- Cannot hurt the seals

Invariably, as every group begins to offer up creative ideas during its 30-minute problem-solving meeting, the solution criteria are used to qualify (i.e., bazooka) every beginning solution. And, the more creative the candidate idea might be, the quicker the criteria are invoked. The team is usually frustrated at the end of the exercise because they can't come up with any ideas that fit all the criteria, despite being able to float plenty of big, high-potential, beginning ideas. This exercise shows teams the inherent dangers of applying criteria too early in the innovation process.

Remember the Other Critical Piece

Everything I've talked about in this chapter is essential to your innovation effort. But even if you have all these elements in place—a great business core purpose, an unchangeable set of core values, an exciting envisioned future with a challenging BHAG/task focus for the innovation effort, as well as clear measures of success—remember that your trip on the innovation highway will lead to a dead end unless your organization overcomes the decision-making issues I discussed in Chapter 6. Good objective setting and decision making go hand-in-hand toward the effective elimination of the Leadership-Empowerment Fable and are issues that must be resolved at the beginning of every innovation effort. Chapter 8 tells you how to eliminate innovation speed bumps, another important step that will help get your journey to innovation off to a good start.

[innovation fuel]

- Understand the difference between telling a team where you want it to go (good) versus telling it what you want it to do (bad).

- Unleash the full power of the objective-setting phase of innovation by helping senior management articulate the critical touchstones for innovation decision making:

  - A clearly-articulated, shared, and motivating strategic vision for the *business* (whether that be the entire corporation, a business unit, a brand, or any other subunit). This takes the form of a core ideology, which includes a core purpose and core values, plus an envisioned future, which includes your BHAG and a vivid description of the future.

  - A clearly defined task focus for the innovation effort

- Make sure everyone agrees on the interpretation of these elements of your strategic vision by thoroughly exploring how team members interpret key words.

- Match the scope of the vision to the level of innovation you want to pursue.

- Determine how you will measure your innovation success before starting but don't constrict thinking by applying the measures prematurely.

- Do not succumb to paradigm confusion by trying to make new ventures adhere to the measurements of success associated with existing lines of business.

Notes

1. James C. Collins and Jerry I. Porras, *Built to Last,* HarperBusiness, New York, 1997, pg. 220.
2. Ibid., pg. 113.
3. Neal Thornberry, "A View about 'Vision'," *European Management Journal,* February 1997, pg. 29.
4. James C. Collins and Jerry I. Porras, *Built to Last,* HarperBusiness, New York, 1997, pg. 93.
5. Ibid., pg. 232.

CHAPTER EIGHT

Smoothing Out Innovation Speed Bumps

Now that you have your core ideology and envisioned future in hand, there is one more step in Phase 1, Setting Objectives, which you will need to complete before you begin Phase 2, Discovery, where:

- You will rediscover (from when you were a child) skills for breakthrough exploration and ideation.
- You will investigate the four marketplace stimulators of innovation (your consumer, your competition, your technologies, and, if relevant, your regulatory environment).

Discovery is where the problems associated with innovation speed bumps (the negative impact of individual behaviors) will start to appear, so this chapter is devoted to identifying the most common speed bumps and helping you learn how to eliminate or at least minimize them. Doing so will have a powerful and immediate impact on your organization's ability to turn new ideas into reality.

But first you must address the individual behaviors that cause speed bumps. This can be accomplished even in the face of strong cultural roadblocks, so a speed-bump-eradication program is a good way to get an innovation effort rolling. The individual learnings that occur as you make speed bumps visible and introduce skills to combat them will provide important building blocks for innovation.

Fostering an environment where new ideas are treated as a valuable commodity will maximize the creativity and energy that people will be willing to invest in the generation and refinement of innovative ideas. This, in turn, will unite these individuals at those critical times when they will need to overcome organizational roadblocks (the tougher, cultural barriers, which inevitably arise to block your path).

Building a speed-bump-reduced atmosphere requires that people learn to recognize the subtle—and sometimes not-so-subtle—individual behaviors that impede innovation. For some speed bumps, building this awareness is all that is needed to significantly reduce the problem-causing behaviors. But for most speed bumps, people also need to learn new habits and skills that will support innovation rather than inhibit it. In both cases, constant vigilance is required to ensure that people don't fall back into bad behavior patterns because, as my Undertoes anecdote in Chapter 2 showed, old habits die hard.

The nine innovation speed bumps that can be found in most organizations and which will be discussed in this book are:

1. Bazookas (both humorous and nonhumorous)
2. Disagreement between words, tones, and nonverbals
3. Bad paraphrasing or no paraphrasing
4. Recognizing the difference between intent and effect
5. Passive listening
6. Rambling
7. Stealing
8. Questions
9. Reactive thinking

This chapter discusses the first eight of these speed bumps, most of which are closely linked. All involve issues of communication, or more aptly, miscommunication. To help you avoid these speed bumps, I will suggest some new and not-so-new behaviors that you and the people in your organization can adopt to avoid these jolts that can slow things down along the innovation highway. The ninth speed bump, reactive thinking, is very closely related to a major innovation roadblock that occurs during Invention. I call the roadblock Newness/Feasibility Schizophrenia, and this roadblock and the reactive thinking speed bump will be discussed together in the chapter on Invention Convergence (Chapter 13).

While this chapter focuses primarily on how these speed bumps disrupt brainstorming, be aware that these behaviors crop up in lots of other types of human interplay. Because personal interaction is the bedrock of successful innovation, you must be aware of how these speed bumps can wreak havoc, not just with brainstorming, but with all the other phases of innovation. Failing to address these speed bumps early on will make your ride through the Dark Night of the Innovator bumpier than it needs to be.

Speed Bump 1: Bazookas (Verbal Warfare)

Bazookas are the most prevalent innovation speed bump. I don't think I've worked with any organization where shooting down ideas was not routine until the far-reaching negative impact of this behavior was pointed out.

Usually, the end result of using a bazooka—particularly a humorous one—is the elimination from consideration of a new idea. This result is often completely unintentional. Most of us don't mean to slaughter ideas with our harmless barbs. In fact, many times we don't even think the idea we've blasted is a bad one. Something about it strikes us as odd, or funny, and we just blurt out our put-down. But just because a bazooka is accidentally fired doesn't mean that damage isn't done. Even friendly fire can kill a good idea.

Depending on what is acceptable behavior in a given corporate culture, bazookas can range from lighthearted jests that elicit chuckles to deadly serious comments that deeply embarrass their targets. Some corporate cultures actually encourage a take-no-prisoners approach during brainstorming. Needless to say, few ideas survive in such a poisonous atmosphere because everyone is so busy waging silly internecine warfare that they fail to notice that the innovation battle has already been lost.

Using a quip to blast someone's idea out of the sky is, unfortunately, almost universally acceptable. And, regrettably, any rebuttal against the bazooka only makes matters worse. Defending oneself against someone's harmless remark gets you labeled as being overly sensitive. Therefore, everyone plays along with the game.

And don't ever forget that the cumulative, behavioral by-product of bazookas is the "revenge syndrome," where at the first opportunity, we engage in and do unto others what has been done to us. We blast their ideas with our own bazookas in retaliation.

The Bazooka Fallout

Beyond the dead idea, the other immediate casualty from a bazooka blast is the person whose offer was shot down. Frequently, that individual shuts down and stops contributing ideas in order to avoid being embarrassed again. Between the bazooka victims who retreat into their shells and the ones who refocus their energy on retaliating against their tormentors, it only takes a few blasts to move a brainstorming session completely off track.

Some individuals love firing bazookas so much that they make this behavior their official role on the innovation team. If you know someone who constantly starts comments with the phrase, "Just to play devil's advocate," you know such a person. Using the it's-a-dirty-job-but-someone-has-to-do-it excuse, devil's advocates justify their negative comments by actually believing that constantly taking the contrarian's point of view helps the group test ideas and prove their mettle.

There is one good thing to say about devil's advocates: they haven't bought into one of the other myths about brainstorming—the false belief that you have to love every idea and can never point out flaws. This myth has evolved because most groups don't understand that brainstorming requires two stages: uncensored ideation (divergent thinking) followed by idea development through open-minded evaluation (convergent thinking). I like to call them the seasons of idea generation.

The problem is not with the notion of testing ideas but rather with how devil's advocates go about doing the testing. My experience is that this type of person has never learned that ideas can be tested without ever resorting to harmful negativity. No one has ever suggested to the devil's advocate that, because nascent ideas are so fragile, you must first gather them in an uncensored, no-bazookas environment. After the gathering, any idea can be thoroughly explored—including its problem areas—without using bazookas. You just have to wait for the appropriate time to do it and you have to use the right (i.e., nonnegative) methods. (See the discussion of open-minded evaluation in Chapter 13 for a description of how to do this.)

Devil's advocates face an important, negative consequence from their behaviors. If you constantly bazooka other people's ideas, you are likely to become the prime victim of the revenge syndrome. This means that people will be more likely to take shots at your ideas, regardless of their merit, even when you don't want to be in the devil's advocate role. Thus, establishing yourself as the corporate devil's advocate can have a huge, negative impact on your ability to win support for your own ideas at almost any time.

The positive, flip side of the revenge syndrome, which devil's advocates also don't understand and almost never benefit from, is that people who are receptive and open to other people's ideas generally find that their own ideas receive a similar, warm welcome.

The worst possible situations occur when the person who is using bazookas is the boss. Nothing destroys a brainstorming session faster than this. It's bad enough if you're worried that your ideas might be shot down by a coworker, but who wants to risk being in the crosshairs of the boss's bazooka? And, in most organizations, nobody's humorous bazookas get a louder laugh than the boss's, which means that the embarrassment experienced by the idea generator can be significantly greater when it's the boss who takes aim at an idea.

Disarming Bazookas

Normally, just calling people's attention to the highly damaging impact that negative comments, even those that are disguised as humor, can have on brainstorming reduces the use of bazookas. Labeling such actions as bazookas goes a long way toward making people think before they make a wisecrack that unintentionally deflates someone else's idea. Setting a ground rule of no bazookas, particularly in the ideation phase of a meeting, gives people a common frame of reference and helps the group work together as a team to discourage bazookas.

You might want to go a step further and fight fire with fire. In the sessions we facilitate, my coworkers and I bring along toy bazookas that shoot small, harmless, soft plastic balls. Everyone is told that if anyone shoots down their idea, they have the right to use the toy bazooka on that person. This playful behavior tends to keep negative comments to a minimum. It also helps defuse those that do happen because calling attention to a verbal bazooka immediately minimizes its impact. Alerting everyone to a bazooka also sometimes takes the focus away from the content of the remark and, instead, emphasizes the fact that the comment was inappropriate and should be ignored.

Why I Use Toy Bazookas

This practice in our meetings began shortly after we started Creative Realities. Even then, I had been using the term *Bazooka Syndrome* for years. In the fall of 1989, we were conducting a new product development project for Bacardi, one of my oldest clients. To our delight, our friend Dr. George Dorion, the chemist I mentioned in Chapter 5, arrived at a session in Boston with a large toy bazooka, which he had assembled at home using three sections of plastic plumbing pipe. With large press-on type he had emblazoned the word *bazooka* along its side.

Needless to say, his gift was a big hit. By the next morning, I had proudly mounted it on the wall of our session room. The Bacardi folks, of course, got a real kick out of that. At the time, I thought that was the end of it.

Two weeks later, a new client group was conducting their first brainstorming session at our office. Despite our every effort and in spite of every facilitation skill I could muster, the president of that small company insisted on pointing out the flaws of every idea that the group was generating. During a break, when he complained that he wasn't seeing anything really new, I pointed out that it was a consequence of his negative behavior.

Nevertheless, after the break, he started doing it again. Suddenly, one of the members of his team got up and removed our Bacardi bazooka from the wall and ceremoniously placed it in that president's hands. Much to the president's embarrassment, the entire group stood and cheered. After the cheering subsided, he started to replace the bazooka on the wall, but was forbidden to do so until he stopped using his verbal bazooka. He was told that, if he stopped, he could pass the bazooka on to the next person who shot a bazooka. He ended up with that toy on his lap for the rest of the day and the session took off. Nobody modeled bazooka behaviors after that. When the session ended, we went right out and bought 500 toy bazookas from a company in Hong Kong and have been reordering ever since.

Some facilitators use rubber balls or similar tossable toys for the same purpose as our toy bazookas. Whatever you use, giving people an accepted, fun and nonverbal way to fight back when someone takes pot shots at their ideas reduces that tendency to criticize and keeps more ideas alive. In addition, such methods help level the playing field, because even if it's the boss who shoots down your idea, you have an authorized way to stand up for your concept. I've even seen very shy people, who wouldn't normally speak out in defense of their ideas, use this physical method to fend off bazooka wielders.

Speed Bump 2: Disagreement between Words, Tones, and Nonverbals (It's Not Always What You Say; It's How You Say It)

Closely tied to bazookas is another form of innovation speed bump—disagreement between words, tones, and nonverbals. We all know that words are not the only way we communicate with each other. Interpersonal communication also includes:

- *Tones.* These define how we speak and cover such things as inflection, volume, anger, hesitation, sarcasm, and whining.
- *Nonverbals.* These are expressed in the body language we display as we speak, including posture and facial expressions. This nonverbal information is processed largely unconsciously.

Studies by psychologist Albert Mehrabian, in the late 1960s, validated the old adage that "it's not *what* you say but *how* you say it." Mehrabian found that nonverbals take the lead in communicating feelings or attitudes toward others. He defined the relative perceptual impact of each element as follows:[1]

Element	Impact on perception of each element
Verbal feeling (words)	7%
Vocal feeling (tone)	38
Facial feeling (nonverbal)	55
	100%

Communication, of course, is great when all of these three elements are in accord. But let one of the dominant elements (tone or nonverbal) differ from what a person is saying, and problems instantly arise. Remember when President George Bush looked at his watch during a presidential town hall debate in 1992? What made the news headlines was not what he said, but rather the image of a president whose body language communicated that he was disengaged from the process. That's one of the better-known examples of what happens when the elements of communication are out of sync.

This understanding that tones and nonverbals can speak louder than words is particularly important during the numerous interactions that take place in the pursuit of innovation. Problems can arise when our interpretations of people's tones and body language are correct *and* when our interpretations are wrong, as they sometimes are. Here are examples of how things can go awry in ways that result in idea killing:

- If someone says he likes your idea but has a bored look on his face, is slumped in a chair, or speaks in a dull tone, you will, most likely, distrust the words. You may be right; he may hate your idea. But it is also possible that the problem may be that this person didn't get enough sleep last night and has no energy. If you misinterpret and start to get defensive, he may, in turn, get irritated because you're not accepting an honest answer. The person may even rethink his original position and decide that your idea is a bad one after all.
- Because people are extremely sensitive about their new ideas, even the smallest negative comment, slightly wrong tone, or unintended gesture can end up shooting down ideas unintentionally.

I remember a class I was teaching in creative problem solving several years ago. In this course, we videotaped the group practicing their skills while working on each other's real problems. In one practice session, the problem owner, who I'll call Fred, was sitting on a couch with Jim, one of his classmates (and coworkers). Early in the meeting, Jim offered up a rather creative wish. In response, Fred emitted a very audible whistle. Not only did Jim totally shut down and check out for the rest of the meeting, but Fred didn't get any more creative ideas from the rest of the group.

During the debriefing of the meeting videotape, Fred was told that he had killed the creativity he had honestly wanted to get on his problem. Before viewing his whistle, he was totally clueless that he had done anything negative. Upon viewing the tape, he was surprised and quite apologetic.

But the story doesn't end there. Despite Fred's apology to Jim and the group, Fred soon experienced the revenge syndrome I mentioned earlier. Two days later, during another practice session, the roles were reversed and the group was working on Jim's problem. During the brainstorming part of the meeting, Fred offered up a pretty nifty wish. Jim immediately turned to the video camera and announced, "Mark, I know I'm not supposed to do this, but I can't resist." He then turned to Fred and neutron bombed his idea to smithereens. Everyone laughed, but the power of the message got through to that group. And, while training tends to heighten people's awareness and sensitivities to these kinds of behavior, do not underestimate the negative consequences of these interpersonal speed bumps on your innovation team in the real world.

As this story shows, tonal and nonverbal rejection of ideas can be just as punishing as verbal ones. The answer to avoiding such debacles is twofold.

First, make people aware of the strong impact that things like facial expressions and posture have on the messages they are both sending and receiving.

Second, people need to become accustomed to checking out their interpretations of nonverbals and tones before jumping to a misleading conclusion. This checking should be done off-line if possible; that is, during a meeting break or at some other time outside the group setting.

Getting one-to-one clarification tends to eliminate defensiveness and other issues that can arise if someone is challenged about a nonverbal or tonal response in front of the group. Group intervention should only be attempted in real time with extreme caution, perhaps with the help of a skilled facilitator.

Speed Bump 3: Bad Paraphrasing or No Paraphrasing (Not Clarifying What Was Said)

One of the best tools you can use to resolve miscommunication problems and make sure everybody is on the same wavelength is paraphrasing. Paraphrasing helps bring out the intended meaning behind words. Even when you think you heard the words correctly, your understanding of the message behind them can be off the mark. So, to increase understanding, summarize the other person's thoughts using different words.

Be aware of the difference between paraphrasing and parrot-phrasing, in which you restate someone's ideas using their exact words. Parrot-phrasing gets you nowhere. By merely repeating the other person's words without establishing that everyone shares the same understanding of their meaning, you falsely give the impression that clear communication has occurred.

Real paraphrasing, on the other hand, not only helps you clarify your understanding of what the other person said but also assures everyone that you are actively listening. If your paraphrase attempt is off base, you have a chance to clarify issues right then, without the danger of moving forward only to realize later that you and the idea offeror are interpreting a concept differently.

Paraphrasing takes some skill. Remember this important rule: A paraphrase is not complete until the person being paraphrased says, "Yes, that's what I mean." The reason this is so critical is that it can sometimes require three or four rounds before the person being paraphrased says yes. An incomplete or incorrect paraphrase can really screw up communication at critical times in any meeting.

Paraphrasing can be a very helpful clarification tool. That is why we tend to paraphrase when we are confused by something. However, it is equally important to paraphrase when we think everyone is in agreement. Too often, this doesn't happen and we proceed along multiple thought tracks instead of one.

Teams that are skilled at paraphrasing—and not parrot-phrasing—suffer from fewer miscommunications. It is not always necessary to check understanding in this way, but in some key situations it is helpful to make sure everyone has a shared understanding before moving forward, especially:

- Prior to accepting an idea for further development or rejecting one.
- Before agreeing or disagreeing with someone's view or information about the problem on which the team is working.
- Prior to judging or evaluating someone's work or actions.

If you are assuming the facilitator's role in a meeting, be careful not to overuse paraphrasing as a summarizing tool. Too-frequent paraphrasing can unintentionally distort others' messages and create distrust in you by making it seem like you are trying to impose your own views or forcing them to defend theirs. Instead, occasionally ask others to paraphrase themselves.

Speed Bump 4: The Difference between Intent and Effect (Not Clarifying Why Something Was Said)

As illustrated in the discussion of tonal and nonverbal communication, the *effect* of doing or saying something can often be unrelated to the *intent* with which it is performed or said. For example, let's go back to my story about Fred and Jim. You probably thought, as Jim and the group did, that Fred's whistle was a put-down of Jim's idea. You assumed that was his intent. However, the part of the story that I didn't share before was that Fred's whistle was in admiration for the creativity of Jim's idea. The intent of his whistle was to convey his like, not his dislike, of Jim's idea. Imagine what a difference it would have made in the outcome of that problem-solving session if Jim (or anyone else in the room) had been skilled enough to check out this misunderstanding!

As with the previous speed bump, people who feel they have been on the receiving end of negative intent need to determine whether their interpretation is correct. For example, if when Fred whistled, Jim had simply asked, "Was that a good whistle or a bazooka?" he wouldn't have been placed in a position of feeling his idea wasn't being appreciated when, in reality, it was very much appreciated. And Fred might have had an opportunity to save his meeting.

Speed Bumps 5 and 6: Passive Listening (Idea Vapor) and Rambling (Yada, Yada, Yada)

Two other speed bumps are linked because they describe a relationship between how we often speak and listen in meetings. We call these speed bumps passive listening and rambling. They arise in almost any meeting scenario.

For example, a group is gathered for an ideation/problem-solving meeting. Everyone is being encouraged to offer up ideas. As soon as someone gets the floor, everyone else tends to listen for a few seconds before dropping out of the conversation and into their own heads. Usually, they will stay there, often preparing a response to, or observation on, whatever the speaker is saying. Many will stay with their own thoughts until they think the speaker is about to finish. When the listeners sense the speaker is nearing the end of his or her floor time, they return to attention because they want to be alert for a point where they can jump in and take the floor themselves. This kind of passive listening is commonly known as the rehearsal because listeners are often preparing their response to the speaker.

Something else can occur as we listen to a speaker. If we're not rehearsing a follow-up to the speaker's point, another of the cardinal sins of innovation may be happening. People are generating lots of ideas without writing them down. This means those ideas will be many times more likely to disappear before being shared. They will become idea vapor. Why? Psychologists have concluded that people can only remember a few thoughts at a time before the memory starts erasing the old data and replacing it with new input. (The debate rages as to whether this number of thoughts is four or up to seven.) Consequently, without a place to store more ideas (like on a pad of paper), we do one of two bad things:

1. We shut down in order to hold onto what we've got, thereby missing any new information.

2. We lose one stored idea for every new one we add to our mental memory cache.

At the same time another unproductive speed bump is being manifested in the meeting: the endless ramble. Few things suck the energy out of a brainstorming session faster than allowing speakers to go on and on when presenting their ideas (yada, yada, yada).

We've all experienced this innovation speed bump. Some people seem genetically incapable of keeping it short. The effect on others in the room is deadly. People lose focus, mentally drop out, feel frustrated, and start looking at the clock. Good ideas that were on the tip of people's tongues get lost as the rambler proceeds to drive an idea around and around and around the block. Side conversations crop up and the next thing you know, the meeting is completely out of control.

Some ramblers ramble because they lack the confidence to just toss out their idea and let it speak for itself. Other ramblers are just so enamored of their own ideas that the possibility that those ideas might not be well received is intolerable. In both cases, ramblers try to make sure that every single aspect of their idea gets out onto the table. They fear that if they don't present all the background and explain in excruciating detail why their idea is so fabulous, others might reject it because they lack some piece of critical information.

The mental mode of the rambler, and most ideators, is the joke-telling mode. We think we need to set up for the punch line. Unfortunately, what ramblers don't understand is that most people in the room have checked out before their punch line gets delivered.

Remedies for Passive Listening and Ramblers

Here are two skills for overcoming these linked speed bumps:

Speed Bump	Behavioral Cure
Passive listening	Active note taking
Rambling	Start with a headline

Active note taking lets you make use of your daydreaming. A pad of paper becomes a repository for every little wisp of an idea that passes through your brain, whether it seems relevant or not. It makes your drop-out time, when you're not listening to the speaker, more productive. It allows you to doodle, something we've all been conditioned not to do since first grade. But doodling can, in fact, be highly useful in creative problem solving, and you should let people know that, in this atmosphere at least, they won't be considered rude if they doodle while someone else is talking.

The headline mode is the opposite of the joke-telling mode, as a way of managing ramblers. It begins with the punch line and is followed by the elaboration. It's the sound bite approach to communication, which is what we've learned from politicians and others who are skilled at getting their messages across in the mass media. Headlines are also increasingly how we deal with an overload of information.

Headlines help people to communicate the guts of their ideas as they begin to speak, when others are most alert to their words. Headlining helps people become skilled at presenting ideas in a way that doesn't stop brainstorming dead in its tracks.

Setting headlines and active note taking as ground rules at the beginning of a meeting helps foster these techniques. Encouraging headlining and active note taking will also create some new dynamics in your meetings:

- Listeners will recognize how fleeting ideas can be and will manage the process more effectively. When something triggers a thought, they will drop out of listening consciously, make key word notes (or a doodle) about the trigger, and then move quickly back into the dialogue.
- Speakers will recognize that listeners may be dropping out for significant periods of time so they will seek to front-load their presentation of ideas with the truly critical information. Listeners can then tune out while speakers are explaining or embellishing and they will still have received most of the important data.

With this pattern, the power of each speaker's offerings will be heard and more ideas will be generated because listeners will take in, capture, and play with more information. Also, more of these ideas will be captured for presentation to the group.

Speed Bump 7: Stealing (Robbing the Idea Bank)

Thomas Edison once declared, "Nobody ever came up with a great idea all by themselves." Most ideas become great ideas when people continue to build and mold a beginning idea until it becomes a truly great idea. Alfred

Marshall, an English economist in the late 19th century, spoke directly to this point:

> The full importance of an epoch-making idea is often not perceived in the generation in which it is made. . . . A new discovery is seldom fully effective for practical purposes till many minor improvements and subsidiary discoveries have gathered themselves around it.

Thus, one of the most important idea-strengthening behaviors people can learn to do in meetings is to build on each other's ideas. To be effective, this building on ideas has to include crediting and acknowledging the people who put forth the beginning idea. This behavior has a tremendously positive effect on the dynamics of the group and encourages people to idea-build throughout the session. But equally important to realize is the fact that the reverse behavior—failing to give credit—can have an extremely negative impact and destroy any trust between colleagues.

If you don't acknowledge the part others are playing in your thinking, you can inadvertently create the feeling that you've somehow taken over or stolen the other person's ideas. Even though the reaction may not be a conscious one, its effect can make the other person reluctant to offer additional ideas and suggestions. By failing to give credit, you can effectively knock others out of the meeting.

Also, you cannot fake idea building. Some people try to say they're building on a previous thought to get their ideas into the conversation or in an attempt to make others think they own a piece of the original idea. This does not work. People feel manipulated and can have a negative reaction if they realize the idea building is not real. If you have to get a thought off your chest, it's better to just offer it rather than try a fake build on another's idea.

Stealing of ideas—no matter how unintentional—also has implications further down the road. If somebody is successful in coopting another person's idea, the chances that the original ideator will cooperate with the innovation effort is greatly reduced. This person, who could have been the idea's greatest champion if given proper credit, may, in fact, become the idea's enemy.

Instead of creating an environment where people hesitate to present ideas for fear that they will be stolen, your objective should be to establish an atmosphere where the great idea is never attributable to just one person. The appropriate crediting behaviors that should be modeled in your organization include:

- Acknowledge others' parts in your own thinking. Statements along the lines of "Ramon, your idea about that made me think about . . ." or "Your point suggests to me that we should . . ."
- Say more than just "Good idea!" Explain why and share what it triggered for you.
- Credit an idea that you disagree with when it triggers another thought for you.
- Create a sense of ownership and, therefore, commitment. If a team member suggests a course of action to you, it is especially important to remember crediting. By making it clear that the other person contributed to your thinking, you allow him or her to have part ownership of the idea. By creating this sense of ownership, you increase the commitment to the overall decisions or courses of action.

Speed Bump 8: Questions (The Well-Intentioned Idea Killers)

This speed bump is often challenged when I bring it up because it contradicts one of the most revered management practices: asking questions.

Questions help a meeting when they provide clarification. But every well-trained facilitator knows that, in most organizations, a lot of questions have nothing to do with eliciting clarification. Instead, they tend to be polite ways of pointing out a flaw in an idea or even killing the concept.

Our experience suggests that over 80 percent of questions in a meeting are not really questions at all. They mask all kinds of other thinking. How many times, after you've offered up an idea in a meeting, does someone ask "What do you think the boss will say?" or "Can we do that within our budget?" I contend that there really is no question in these questions. They are masked rejections or masked ideas.

Questions are key climate indicators. When I work with a group that tends to preface everything with a question, I know that there is little trust, few opportunities for real creativity, and probably not a lot of healthy humor (which is critical to innovation and is the opposite of the humorous bazooka).

Here is a powerful illustration of the destructive power of masked questions: A sales manager of a large industrial manufacturing company told Yves, one of his sales people, to unload the inventory of frangis (a made-up word), one of their products that had been sitting in the warehouse for too long.

About a week later that sales manager received a phone call from a prospect he'd been wooing for a long time. That prospect had been short-shipped on frangis by his regular supplier and needed help to keep his manufacturing lines running.

This was the sales manager's big chance with this prospect. Excitedly, he hurried down to Yves' office and asked, "Did you sell the frangis that I told you to sell?"

Yves had not sold the frangis. But out of fear because he didn't know what was behind the question (why it was being asked), he told the boss that he had sold the frangis. The disspirited sales manager returned to his office to call his disappointed prospect.

Now, you might choose to debate whether Yves did the ethical thing by lying to the boss, but situations like this happen every day and they are often caused by questions that get in the way of communication.

There are two ways to address this very destructive innovation speed bump depending on whether you are the question asker or the question recipient:

1. If you are the question asker, support your question with an explanation of why you are asking it. Imagine if the sales manager had said, "Yves, did you sell the frangis that I told you to sell? Because if you haven't, it would really help me with a hot prospect that I've been pursuing for years." He might have learned the truth and opened a customer door.

2. If you are the question recipient, probe the reason for the question. For example, what if Yves had said, "Gee, boss, I'm waiting for a call from someone who I think wants them. But why are you asking?" He might have shared in his boss's success.

Smoothing Out Speed Bumps Is Only the Beginning

This chapter has introduced many of the innovation speed bumps that can slow any innovation effort. Make no mistake, giving people the awareness and tools they need to individually support innovation can make a significant difference in your organization's ability to achieve it. In the chapters ahead, you will begin to appreciate how the simple yet powerful individual skills discussed in this chapter can provide a strong foundation for the more complex efforts you will need to employ to overcome the group-based, cultural barriers (roadblocks) that are likely to arise in the path of your team.

[i n n o v a t i o n   f u e l]

- Provide training that teaches team members the dangers of innovation speed bumps (individual behaviors) and how to overcome them:

    - *Bazookas.* Don't tolerate any attempt to shoot down other people's ideas, even if it is disguised as humor.

    - *Challenges positioned as coming from a devil's advocate.* Teach individuals who want to take a devil's advocate type of approach that, during divergent ideation, it is not the right time for testing ideas.

    - *Disagreement between words, tones, and nonverbals.* Make sure team members understand that how you say something (tones) and how you look when you're saying it (nonverbals) have far more impact than the words you actually say. The same is true when you're just listening; body language speaks volumes.

    - *Bad paraphrasing or no paraphrasing.* Help people learn the value of good paraphrasing in establishing a common understanding. Good paraphrasing is much different from parrot-phrasing, in which you merely repeat what the other person says word for word without really connecting with the true meaning of those words.

    - *The difference between intent and effect.* Encourage your team to check in with each other about the intent of comments that have a negative effect on them. It could be they will often find that the intent of the other individual was, in fact, not negative.

    - *Passive listening.* Provide people with writing materials and make frequent reminders that note taking and doodling are helpful to ideation.

- *Rambling.* Teach the skill of headlining ideas followed by elaboration. Enforce this concept when someone starts to ramble.

- *Stealing.* Make sure people credit others when they build on their ideas; educate them in the proper ways to do this.

- *Questions.* If a question is asked, immediately probe for the idea behind it. Promote the notion that it's better to just present your idea rather than to disguise it as a question. If you must ask a question, at least support it with an explanation of why you are asking it.

- As part of your speed-bump-eradication program, provide playful (i.e., nonthreatening and fun) ways for people to point out and resolve individual behaviors that are slowing progress (like a toy bazooka).

- Be aware that smoothing out speed bumps requires constant vigilance by all team members of both their own behavior and that of others. This is true no matter how much experience you and your team have; everyone is capable of slipping back into old, unproductive habits.

Note

1. Albert Mehrabian, *Silent Messages,* Wadsworth Publishing Company, Inc., Belmont, CA, 1971, pg. 44.

[PHASE TWO]

Discovery

CHAPTER NINE

Gearing Up for Discovery

Now you've established the strategic objectives for your innovation effort and put processes in place to overcome the Leadership-Empowerment Fable. You've provided role definition for Championed Teamwork decision making and truly stretched your definition of a cross-functional innovation team. Under the traditional model of innovation, which most companies follow, you would now jump right into the ideation process. But, as described in Chapter 4, under the new, improved flow of innovation that I'm recommending, you will now enter an exciting exploratory phase that supports the pursuit of breakthrough innovation. This is the Discovery phase, and I think it is the most interesting and exciting part of the whole innovation process. I guarantee that you will be amazed and delighted by the richness of the team learnings that can occur in Discovery.

For me, the Discovery phase is analogous to the due diligence phase in the business acquisition process. Due diligence provides time for you to delve into the nooks and crannies of your candidate acquisition, to kick tires and challenge the apparent truths. Our Discovery phase is designed to help the innovation team accomplish exactly the same thing.

As mentioned before, many companies skip the Discovery phase or don't even know that one exists. As you can imagine, this works about as well as skipping due diligence when you're buying another company. Do not delude yourself into believing you already know everything you need to know about thinking-outside-the-box skills, about your consumer, your competition, the regulatory environment, and relevant technologies. Once you become involved in the types of learning that emerge during Discovery, you will be amazed at how much you didn't know about the key stimulators of innovation.

Activities undertaken in the Discovery phase lay the informational foundation for everything that follows in your innovation effort. Before moving into Discovery, you first gear up by making sure your team members have the skills they will need for effective brainstorming and creative problem solving; skills that are critical to every phase of innovation. Once armed with these skills, you will examine your innovation task from all sides so that your team members can begin to make new mental connections and associations of thought that will produce breakthrough ideas. Think of Discovery as filling your minds with the pieces of the puzzle that will eventually come together to form breakthrough new ideas in Phase 3, Invention.

The two key success factors for Discovery are:

1. *Well-prepared minds.* These are capable of valuing and mining the absurdity of new ideas.
2. *A willingness and methodologies for exploring the four key drivers of innovation:*
   - Consumers
   - Technology
   - Competition
   - Regulatory environment

Preparing Minds for Discovery

In addition to the due diligence analogy, I use a second analogy to describe the goals of the Discovery phase because Discovery is the least results-oriented phase in innovation. I tell the teams I work with that in Discovery we're going to adopt the old Bell Labs mentality. Before the breakup of AT&T, Bell Labs was treated like a pure research arm. Their charge was to simply create new possibilities. They were encouraged not to worry about cost factors, marketability, or potential profitability. All they needed to do was invent something new; the rest of AT&T would figure out what to do with it.

That pure research mentality is what the Discovery phase is all about. This approach has a not-so-hidden agenda. All of my experience shows that by removing the requirement for tangible results from this phase you actually enable people to free their minds to come up with exciting break-through possibilities. The irony is that more breakthrough thinking happens in Discovery precisely because it is not expected from this phase.

The reason why eliminating the demand for tangible results works so well is the paradox of relaxed concentration that I alluded to in Chapter 1, in the story of how Archimedes discovered the law of physics for liquid displacement. Albert Einstein referred to this phenomenon when he asked, "How come I get my best ideas when I'm shaving?"

Trying to learn more about relaxed concentration, my colleagues and I frequently ask members of our client teams when they typically get their best ideas. Not surprisingly, I don't recall anyone ever saying they get their best ideas while sitting at their desk during the normal work day. Instead, the common theme of their responses mirrors Einstein's query. In a state of relaxed concentration, we tend to lower our mental censors and play with new possibilities more open-mindedly.

Allowing your innovation team to not be obsessed with tangible output until Phase 3, Invention, and giving them permission to relax and have fun during Discovery, usually results in some of the richest, most breakthrough thinking. This requires providing some training in skills for breakthrough thinking and creative problem solving as Discovery begins.

The Talent Assumption Roadblock

Making sure the people on your innovation team have the necessary creative thinking skills may seem obvious yet it is a step that I find companies strongly resist. I'm constantly told, "We've got really smart people, and we don't have time to teach skills."

This brings up the first innovation roadblock you will encounter in the Discovery phase. I call it the Talent Assumption. This refers to the widely held belief that when it comes to thinking creatively, some have it and some don't. This is only partially true and the consequences of this cultural roadblock can seriously deprive any company of the full potential of its people.

It would be foolish to not agree that some people are more creative than others. But any well-trained facilitator will take issue with the some-don't-have-it thinking that lies behind the Talent Assumption roadblock. To get to what I believe is the heart of this matter, I need to first review how many of us have come to believe that some people aren't capable of being creative.

The Left Brain–Right Brain Metaphor

The terms *left brain* and *right brain* are widely used to describe the two mental selves that we each possess. These descriptors for how we think have evolved from scientific research conducted in the late 1960s with patients who suffered from epilepsy. The results of that study hypothesized that each hemisphere of the brain controls different behaviors and thought processes. It concluded that the left hemisphere is where skills like logic and mathematics reside while the right hemisphere is where emotions and creativity reside.

Throughout the 1970s and 1980s, further biological research showed that this theory was full of holes. Science now knows that the brain is far more complex than can be explained with a hemispheric model. Yet the metaphor of the left brain and the right brain continues to be used even today to explain this important paradox in human behavior. People still commonly put a label on someone else when they want to indicate whether they feel that person is creative (i.e., right-brained) or not creative (i.e., left-brained).

One reason why we cling to an outdated theory about how our brain functions lies in the field of psychology, which acknowledges two paradoxical sides of our nature. George Prince calls these sides the "safekeeping self" and the "experimental self." Each is governed by a totally separate set of goals and decision-making criteria:

Safekeeping Self	Experimental Self
Reassures	Loves to take risks
Evaluates reactively	Breaks rules
Is realistic	Speculates constantly
Is logical	Sees the fun in ambiguity
Hates surprises	Makes impossible wishes
Loathes risk	Is intuitive
Punishes mistakes	Values irrelevance

Here's the catch when it comes to being creative. Most of us tend to use our safekeeping skills to think creatively. Now, how can that possibly work?

Staying safe, while being creative, is surely a paradoxical challenge that no one can satisfy.

The reason the safekeeping self is so dominant for most people is that from the age of six our entire lives are dedicated to refining and honing our safekeeping skills. It begins in first grade when we start to play "guess what's in teacher's head." Then comes puberty and high school, with our painful need to fit in. High school is followed by college, where we are again taught to retain facts, value logic, and follow the rules. As MIT's Media Lab professor Mitchel Resnick points out, "In kindergarten and graduate school [the two most unstructured educational environments], people are encouraged to explore and create. In between, they spend large amounts of time filling out work sheets and listening to teachers."[1]

Schooling is followed by either military service or a job, and we know how creative those environments encourage us to be!

In all of these life phases, who encourages us to think like a five-year-old child, whose experimental self is still thriving? Where are we taught and rewarded to get back to that wide-eyed innocence that enables us to explore the world with naïve fascination? The answers, unfortunately, are no one and nowhere.

Functional Biases

The sad fact is that we all—even the most creative among us—tend to lose those wonderful traits of the experimental self that we possessed in abundance during early childhood as we grow older. Contrary to the Talent Assumption misperception that pervades most organizations, we all have vastly more creative capacity than we ever expose to the world, especially in the workplace. And this tendency is further exacerbated by functional bias.

For example, we too often expect the creativity in our companies to reside in certain functional silos like marketing and R&D. And, even in these places, the experimental self can run into trouble when competitive pressures send a shudder through senior management. Mitchel Resnick also points out that in recent years many research labs—traditionally the places where people have been most free to speculate, break the rules, take risks, and to just generally play—have been told to "sober up and get serious about product development."[2]

Few of the other functional areas are expected to be creative at all. Manufacturing tends to be perceived as being fairly rigid and set in their ways. Finance and legal departments are for counting beans and saying no. (I once had a corporate lawyer introduce himself in a technology exploration session with this observation: "Hi, I'm in the legal department and during the last year I think I've said no to just about everyone in this room . . . at least once.")

If you believe in the power of cross-functional teams, wouldn't it be wise to unleash the creative potential in every member with an orientation to some skills for creative thinking and problem solving? You won't really be teaching new skills. Your team will just be relearning the ones they used as kids. If you don't provide for this relearning, you will reinforce the Talent Assumption roadblock. If you do introduce these skills, you will have a team that is vastly better prepared to discover Farson's invisible obvious as they explore the four key stimulators of innovation.

Innovation's Two Critical Thinking Skills

Look to classics like Roger Von Oech's *A Whack on the Side of the Head*, George Prince's *Practice of Creativity: A Manual for Dynamic Group Problem-solving*, Mike Vance's *Think Out of the Box* and *Adventures in Creative Thinking*, or anything by Edward DeBono for plenty of guidance on how to improve your team's creative thinking skills.

The two important creative thinking skills that you should try to build are:

1. Connection making
2. Open-minded evaluation (reflective evaluation)

I'll discuss open-minded evaluation in Chapter 13 because it is very closely tied to and becomes most critical to success in the Invention phase. The other skill, connection making, is where breakthrough thinking begins. Quite simply, connection making is the term that describes the skill of connecting or mentally linking seemingly unrelated thoughts in a new way that sparks a breakthrough idea. Archimedes' overflowing bathtub, which led to the displacement of liquids law of physics, as described in Chapter 1, is one example of a famous idea that was the result of a breakthrough mental connection. Here are a couple of others:

- *Moveable type.* This resulted from Guttenberg's mental connection between children playing one evening with outdoor lanterns (each moving independently and floating in air) and the concept of moveable type.
- *Velcro.* Swiss inventor George de Mastral took his dog for a walk and when they returned home, he noticed that the dog's coat and his own pants were covered with cockleburs. His inventor's curiosity led him to put the burs under a microscope where he saw the natural hook-like feature that made the burs stick to the soft loops of the fabric of his pants and of the dog's fur. The connections this information made in de Mastral's mind led to the idea of creating a closure that combined hooks and loops.

There is no real secret to brilliant connection making, beyond the willingness of the human mind to play with a mental stimulus that, at first, appears to be irrelevant or silly. This is why the Talent Assumption roadblock is such a thinking killer because it never allows for the possibility that everyday people might be capable of thinking creatively by making connections.

A Recent World-Class Connection-Making Example

The wonderful thing about breakthrough connections is that they can happen in the unlikeliest of situations. As painter Grant Wood said, "All the really good ideas I ever had came to me while I was milking a cow." Here's one that happened to a client.

Adrian Crisan is a software engineer at Compaq Computers who participated in an innovation program we conducted for Compaq's Commercial Desktop division a couple of years ago. As part of its core team, Adrian went through our basic skills training in connection making and open-minded evaluation at the start of the project.

During the Invention phase, we held a brainstorming session at a wonderful off-site facility in Brenham, Texas. It was the perfect place for brainstorming because, at 90 miles from their headquarters in Houston, it was very quiet and remote, a great place for people to get away and concentrate on new ideas.

At the time, the facility's main meeting room had only one electrical outlet with two plugs. Unfortunately, for all the computer laptoppers in this session, our on-site technographer, who transcribes the output of the session on computer, needed to use both of them for his laptop and printer. This forced all of the laptop techies in attendance to wait until the end of the day before they could connect (literally) to the outside world because they could only get a couple of hours of power out of their laptop batteries. This was great for our session because it kept people focused on the here and now, but it was a bit frustrating for the laptoppers. This inability to connect created an innovation opportunity—an unmet need.

Here's where the connection-making breakthrough happened. It seems that Adrian began to daydream as he drove the long, flat, boring 90 miles back to Houston after the session ended. His mind began to jump back and forth between his frustration with a lack of electrical power for his laptop and the fingers of our typist. This led him to associate, or connect, the powerful action of fingers typing with the movement of rotors in an electricity-generating power plant. That led him to a wish that he could power a laptop with his finger power.

This was a world-class connection. And, like so many great ideas, it began in the apparent world of absurdity. But, because he knew that beginning ideas like this can be played with and refined, he experimented with it and developed a patentable concept for powering a laptop from finger power.

Now, one might argue that Adrian Crisan is simply a breakthrough thinker who would have come up with that idea anyway. Nevertheless, during interviews with industry journalists, he chose to acknowledge the skills he learned in our training for providing him with the mental tools that allowed him to be receptive to the seeming absurdity in that beginning idea.

The point here is that everyone is capable of achieving higher levels of creative thinking if their organizational culture supports it and if the individuals are trained in some helpful skills.

Look Ahead, Not Behind

Another cultural stumbling block that arises in Discovery is the tendency to predict the future based on the past. As you start to gather data, you're likely to come across evidence of past failures at achieving newness. Such information can seriously dampen enthusiasm as you move forward. People start to limit their thinking based on a fear of future failure. To avoid this problem, people must be constantly reminded of one simple ground rule: That was then; this is now.

Just as a mutual fund prospectus constantly stresses the warning that past performance is no indication of future performance, so too is a company's past performance with innovation not something on which to base judgments about future performance. Organizations and cultures can change. People can learn new skills. Innovation roadblocks and speed bumps can be minimized.

Nuggets—The Discovery Sound Bite

Here's one more thing you need to know as you prepare for Discovery. If you do a good job in this phase, you will assemble a huge pile of new and exciting information about the four drivers of innovation. In fact, don't be surprised if you amass so much data that you may feel overwhelmed by it. At the same time, you may be so stimulated by everything you've learned that you feel compelled to share every single fact and figure with everyone who will be participating in the Phase 3 brainstorming. You must fight this urge.

Instead, understand the critical difference between providing people with raw, unfiltered data and giving them useful insights that you've distilled from the information you've gathered. When Discovery is complete, doing a data dump of all that you've learned may make the brainstorming group, some of whom will not have been part of the Discovery work, feel unable to take it all in. This produces the Information Overload roadblock. When people suffer from information overload, they focus their energy on trying to master all of it. This can be very enervating just at a time when you want people's mental energy to be at its highest for brainstorming.

As Nobel laureate Herbert Simon pointed out in 1995, attention, as a commodity, is in short supply these days for three reasons: First, a day has only 24 hours. Second, the human capacity to pay attention is limited. And third, people are inundated with information.[3] Given these conditions, the last thing you want to do is hand people a thick binder full of new data that they will need to grapple with as part of your brainstorming process.

Instead, what you will need to do during Discovery is to distill your learnings into only the truly meaningful essences. Then, during Phase 3, Invention, you will be ready to share with people only the information that is truly new and insightful, which will provide new perspectives that can lead to breakthrough ideas.

Here's how to do this. For each of the four drivers of innovation that you explore in Discovery, you should narrow the material you present in your brainstorming sessions down to no more than 10 to 15 insights or thought stimulators (maximum 50 total). We call them *nuggets* because they encapsulate a lot of thinking. The learnings you choose to share in these nuggets, of course, should be those that have the highest likelihood of stimulating new thoughts.

Now, you're probably saying, "But what about all those good ideas we come up with in Discovery that will get left on the cutting room floor?" Don't worry. These ideas will be offered up at later times, when appropriate, by many of your Discovery participants who should also be invited to participate in your brainstorming. You can count on the fact that people will champion those ideas that they really like, even if they aren't included in your 50 nuggets.

Think of this process as preparing the sound bites from your Discovery work. At the end of each module of the Discovery phase, sit down and ask yourself what were the ten most important things you learned. At my company, we present nuggets in the form of a headline and a bright, colorful, cartoon-like drawing. We work with a number of artists, including my friend and fellow Colgate Thirteener (college singing group) Rod Thomas, who manages to catch the paradox, humor, or irony that makes a nugget human. A picture truly is worth a thousand words and if the headline is brief and catchy, the whole thing will stick in people's heads. Next thing you know, they will begin to form ideas around it and make unexpected connections between nuggets. Instead of feeling like they are drowning in data, they will soon be overwhelming you with new ideas.

Following the Innovation Fuel for this chapter are some examples of nuggets (created by my friend Rod Thomas[4]) from some work we recently completed, including a couple from Reichhold, a North Carolina firm. Reichhold is a business-to-business developer of specialty chemical products. Use your imagination to envision these black and white versions in lively, Sunday comics–like color, which really does help stimulate creative thought.

[i n n o v a t i o n   f u e l]

- Bring a pure research mentality to the Discovery phase. By not imposing a requirement for tangible results, you will encourage people to fully explore all possibilities. This will produce the richest possible array of learnings to bring into the Invention phase.

- Provide training that helps people become proficient in connection-making skills and in open-minded evaluation of ideas.

- Avoid the fallacy of the Talent Assumption. Understand that, with proper training in creative thinking and problem solving, people from all functional areas of your organization can be counted on to offer and build upon good ideas.

- Understand that the left brain–right brain theory of brain function is outdated. Nevertheless, the psychological metaphor of the safe-keeping self and the experimental self can be used to build an environment in which people feel safe to offer up their ideas.

- Appreciate the critical difference between providing people with raw, unfiltered data and giving them useful insights from the Discovery phase.

- Provide Invention team members with a limited number of key insights (nuggets) focusing only on those Discovery learnings that are most exciting and most likely to stimulate truly new ideas. Present no more than 10 to 15 of these nuggets for each of the four drivers of innovation.

- Present your nuggets in a creative, stimulating format. A bright colorful, cartoon-like drawing with an insightful headline is one way to capture your team's imaginations.

Notes

1. John Yemma, "All Play and No Work," *Boston Globe,* July 12, 1998.
2. Ibid.
3. Don Tapscott, David Ticoll, and Alex Lowy, "Relationships Rule," *Business 2.0,* May 2000, pg. 315.
4. Artwork used by permission of the artist: Rod Thomas, 16 Grasmere Road, Needham, MA 02494, 781-449-0480.

[FIGURE 9.1] Nugget 1: Eat This

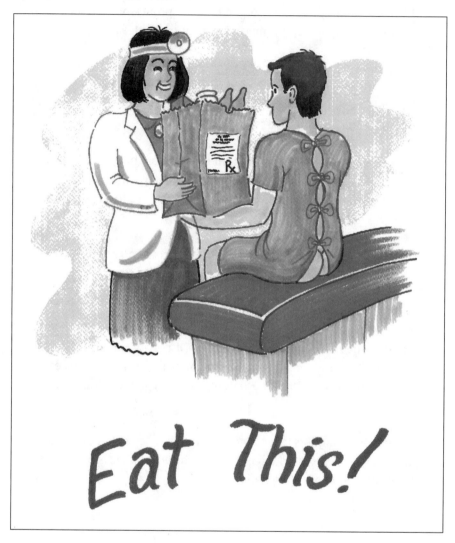

Just as we heard about the potential impact of the Internet years before it became a reality, so too are the food and pharmaceutical fields aware that the marriage of food and medicine (called nutraceuticals) is just over the horizon. This Discovery phase nugget, with its visual suggestion that a doctor might someday soon prescribe food products to address medical ills, helped an over-the-counter medicine client significantly broaden their innovation vision.

[FIGURE 9.2] Nugget 2: Safe Adventure

I've used this trend nugget with numerous clients to discuss a social paradox that affects many industries. Just like these two guys whitewater rafting in a bathtub, many consumers want to experience new thrills without any of the dangers or inconveniences usually associated with adventure. This has both literal and analogous implications. Quite literally, its impact can be observed in the explosion of almost risk-free travel adventures ("I want to go whitewater rafting on the Colorado River, but I want somebody to cook all the meals and make sure I don't get hurt."). On the analogous side, it manifests itself in ways that range from the growth of spicy, ethnic foods to day trading on the Internet.

[FIGURE 9.3] Nugget 3: Small Indulgences

Here is a trend nugget that applies across many industries. It describes how people nowadays reward themselves in many small ways that make life more enjoyable and provide instant gratification, regardless of their socioeconomic status. This trend speaks to the fact that not all of life's pleasures are affordable, even in a booming economy, but we can all afford some small indulgences. For example, you may not be able to afford a Mercedes, but you can buy a Dove Bar and enjoy a rich, tasty treat. Starbucks is one organization that has built a thriving business on providing everyone with the potential for an everyday small indulgence.

[FIGURE 9.4] Nugget 4: Clean Your Kitchen

This nugget acknowledged a beverage company's core competency in tropical fruit and encouraged them to think about how they could use that competency for other purposes. Cleaning your kitchen with tropical fruits speaks to the astringent quality of some tropical fruit juices. Nuggets like this point to the need to push the envelope in terms of defining core competencies and then drive ideas from that whole range of capabilities, not just the ones that come readily to mind at first glance.

[FIGURE 9.5] Nugget 5: Kids Are the Change Agents

The old pattern in which parents taught children has been reversed in many instances when it comes to technology in the home. While many adults still have VCRs with clocks that blink 12:00, their kids are downloading songs from Napster and burning their own CDs. This shift has big implications for product design, service delivery, and marketing strategies, and this nugget drove ideation around this change.

[FIGURE 9.6] Nugget 6: Food Goes to Finishing School

This trend nugget describes the increased sophistication of the American palate, which has moved far beyond the meat and potatoes culture that epitomized American cuisine after World War II. We now bring a far wider range of foods from all parts of the world into our homes and are accustomed to being served intricate recipes, with lots of ingredients, when we go out to dine. Companies in all sectors of the food industry can use this nugget to stimulate ideas about how to serve the increasingly upscale taste buds of consumers, from fast food to white table cloth restaurants, and in supermarket and specialty retail outlets.

[FIGURE 9.7] Nugget 7: Graying of America

The trend illustrated in this nugget is not a huge revelation. But, as the art-work shows, manufacturers and marketers need to realize that today's senior citizens and the baby boom generation that is rapidly approaching retire-ment age are completely different from earlier generations of seniors. They are healthy and have plenty of disposable income, and many are entering into second and third careers. All these changes have huge business implications that can drive innovation.

[FIGURE 9.8] Nugget 8: The Baloney Generation

Here's another major trend that is having a big impact on middle-aged Americans sandwiched between parents, who are living longer, and their children, who are living at home longer (or returning home to live more often) than in previous generations. This nugget helps companies that are developing products and services for the baloney generation to come up with ideas on how they can help ease the stress of being caught between the conflicting needs of elderly parents and children.

[FIGURE 9.9] Nugget 9: Most Convenient Appliance

This nugget helped a client avoid the common problem of defining competition too narrowly. The learning—that the easiest source of food for many people is the drive-through window—applies to both food companies and appliance manufacturers. This awareness helped this innovation team to think about competition in a totally new way during their ideation phase, which resulted in a broader range of creative, new ideas.

[FIGURE 9.10] Nugget 10: Smart Trucks

Smart Trucks

During a Discovery session on technology, one division of a food conglomerate learned that another division used technology similar to the tracking computers used by overnight delivery personnel to monitor in-store inventories of perishable products. This no-brainer linkage of people from different divisions, who otherwise would never have met, resulted in the simple transfer of this capability across internal silos in a way that helped drive new ideas for our client's division.

[FIGURE 9.11] Nugget 11: The Right Flower

Nuggets can be used to help develop your strategic vision just as they are used in helping to drive discovery or invention. Because reorganization was in the wind, this nugget helped our multidivisional client to begin to explore the best ways to allocate its various functional resources.

[FIGURE 9.12] Nugget 12: Curiosity Killed the Cat

This nugget was developed during our Reichhold strategic visioning program. It helped the company make sure it addressed the need for making continuous innovation part of its core values, so that it would avoid the problems experienced by companies that rest on their laurels. The implication may appear to be obvious (that complacency is innovation's worst enemy), but the therefores, in terms of strategic innovation practices, can be profound.

[FIGURE 9.13] Nugget 13: Greater Concern for the Environment

A client in the industrial manufacturing business developed this nugget about ten years ago to stimulate awareness of the growing concern for the environment. While not earthshaking in terms of the trend's newness, it was a good sensitizer to remind the innovation team to take environmental concerns into account during ideation. It could be used today across a wide range of industries that need to make sure to be environmentally conscious when they innovate.

[FIGURE 9.14] Nugget 14: Siren Packaging

This Discovery nugget explores the use of packaging as a seducer. It generated ideas about what can be done visually and with aromas to attract buyers. The interesting thing here is that this nugget could be used to drive ideation with a whole range of packaged goods, such as health and beauty aids, food, and cleaning products (both laundry and household-related).

[FIGURE 9.15] Nugget 15: It's Not What It Costs

As the economy began to improve in the mid-1990s, people became more willing to pay for things that made their lives easier. In an age where time is the commodity in shortest supply, it's no longer just the rich who see value in something that goes beyond the nature of the product or service itself to something deeper, more psychologically driven. Not surprisingly, some people are also still hooked on price as the driver of their purchases, so warehouse stores have grown to be popular, too. This nugget encourages ideators to look at these price–value paradoxes and think how they apply to their markets.

CHAPTER TEN

Discovery (Part 1)— Solving the Consumer Conundrum

In Chapter 9, I suggested that there are four stimulators of innovation, each of which needs to be fully explored during Discovery. They are:

1. Consumers
2. Technology
3. Competition
4. Regulatory environment

This chapter covers consumer exploration (the Consumer Conundrum that I discussed in Chapter 3—to really satisfy consumers by leaping beyond them). The remaining stimulators (technology, competition, and regulation) are covered in Chapter 11.

Uncovering Unarticulated Desires

Note that I am not prescribing that you meet consumers' unmet needs. Although this is what most marketers talk about these days, this is actually a far too narrow definition of what you need to achieve. The truth is that very few of us have unmet needs. For example, if you look at where most Americans fall on Abraham Maslow's classic hierarchy of needs, it becomes clear pretty quickly that we have progressed far beyond the first two levels: physiological needs (food, water, oxygen, etc.) and safety needs (security, stability, order, freedom from fear, anxiety, and chaos).

Instead, in most industrialized countries, our 21st-century economies focus primarily on fulfilling people's wants and desires. For instance, I remember hearing the results of a research report many years ago on the reasons people initially bought home computers. The top reason selected (at a very high level of statistical significance) was because they wanted one.

In such a world, you can only truly delight consumers—and grab market leadership—with breakthrough innovations that satisfy unarticulated desires. This is harder to do than to uncover and meet a basic life need. But it can be done. It just requires some subtlety and educated interpretation.

For instance, we like to use a consumer-probing technique that gets at hidden desires. Instead of asking consumers what they are dissatisfied with in the products they use, we now ask consumers, "What are you putting up with?" This is a more powerful stimulator of new opportunity areas because this question elicits answers that get at what may only be an unconscious, unstated annoyance. Yet, it may provide a meaningful opportunity for innovation.

Here's a personal example that gets at what I'm talking about here. Last year, I tried the Atkins high-protein/low-carbohydrate diet, which allows for bacon and egg breakfasts. Unfortunately, my wife, Amy, was going crazy over the destruction of the kitchen from my daily bacon frying. So I tried to broil it. It greased up the oven! Then I tried to nuke it in the microwave. It greased that up too! So I settled into a pattern of kitchen cleanup that was time-consuming but made Amy happy.

One day, with great delight, she returned home from the supermarket with a wonderful new package of Hormel's microwaveable bacon: individual plastic bags with four strips of bacon each and a grease-absorbing towel. Just nuke, open, and enjoy. No grease, no mess.

Now, if someone had stopped either one of us for some in-store market research, I don't think we would have mentioned bacon preparation if asked what we were dissatisfied with in the kitchen. But what if we were asked what we were putting up with in the kitchen? I think both of us would have quickly mentioned the bacon. This is the kind of subtlety you must use in probing your consumers.

Clearly, it's not easy getting a handle on consumer motivations, desires, and purchase practices. Often they say one thing and do the opposite. This is especially true when discussing topics like personal finances or nutrition. People talk about money logically, yet their spending patterns often defy logic. And, as one of my consumer product food clients likes to say, "They talk thin, but they eat fat!"

How to Turn Consumer *Input* into Consumer *Insight*

Having accepted the need to lead consumers rather than follow them, how do you do that? In other words, how do you skate to where your puck will be two to three years into the unknown future?

Here are four ways to gather invaluable information that will help you successfully anticipate (make an educated gamble on) how your consumers will spend their money in the future:

1. *Survey the landscape for consumer trends.* Look especially for ones that may not, on the surface, appear to be directly related to your industry. Broad demographic and lifestyle developments as well as general business trends may hold clues about what your consumers will soon expect from your company. Information from a wide variety of trend sources is usually needed to provide this grist for your innovation mill.

2. *Talk directly to consumers and prospects about their wants and desires.* To do this well, such dialogues need to take place on an ongoing basis because what people want and expect from you is constantly changing. Also, your consumer-focused activities must recognize that a one-size-fits-all approach will not work. Explore the subtle differences between market segments and use this knowledge as the basis for your innovation efforts. In fact, as has often been said recently, the truly great innovators of the future will be those who are able to customize their service offerings to segments of one!

3. *Immerse yourself, physically, in the marketplace.* Spend time out where the rubber meets the road—wherever people buy your goods and services and put them to use. Observations you make in visiting the real world and in probing how you are seen through the eyes of consumers and the people who influence their decision making can have a dramatic impact on your thinking (as examples given later will show).

4. *Examine your own technology infrastructure and consider technologies that are on the horizon.* Technology may provide the answers to many of the

wants and desires your consumers have, so you need to keep abreast of what new developments are coming even if they aren't immediately ready for commercial use. I will talk more about how to explore the possibilities of technology in Chapter 11.

Bringing these four elements together will enable you to develop your best-educated assumptions about where consumers are going to be down the road. Then you can innovate toward that target. In this way, you will create innovations that consumers will want, resulting in strong relationships that withstand the attack of competitive forces.

Trend Watching

Hopefully, you are on top of trends within your own industry. But are you constantly on the alert for lifestyle trends or new developments in other businesses that might impact your business somewhere down the road? And do you really mine trends from all these areas for fresh ideas about what consumers will expect from you in a few years?

Here's an example. In the mid-1980s, I was flying to a brainstorming session with a manufacturer of small appliances. In a copy of *USA Today* that the flight attendant had given me, I came across a factoid that stated something like this: By the year 2000, there will be X times more domestic cats than dogs (I don't remember the exact number). I decided to throw this trend into the mix as one of 20 we would present to the group.

Their first reaction, of course, was total bewilderment as to what pet ownership had to do with small, household kitchen appliances. Undaunted, I asked about the emerging lifestyle changes that might drive this trend. They started talking, tentatively, about their personal experiences with pet ownership. Soon it became clear that cats require less attention than dogs (e.g., they don't need daily walks).

From there, the group began to discuss the evolution of the family, particularly the growth in two-career households. Eventually, this led to a linkage between the work trend and the convenience of cat ownership over dog ownership. Finally, the conversation began to focus on how the changing patterns of family life could impact cooking habits in the home. Ultimately, a discussion that began with a trend that at first seemed to have absolutely nothing to do with small appliances evolved into a highly productive brainstorming session that produced far more innovative ideas than if we had just stuck with discussing obviously relevant trends in the household appliance field.

The lesson here is to cast a wide net for trends that can be used to take your brainstorming into new, broader directions. The result, most often, will be truly new concepts that would not have arisen by sticking strictly to the easily identifiable trends within your own industry.

But, you ask, how can I know which forecasts will be the right ones? The answer was best expressed by a colleague who was asked this same question by a client as we were about to share several trends at the beginning of an invention session. This colleague encouraged the group to imagine that trend implications and forecasts are like the pellets from an elephant gun. Many of them miss their mark, but they still can bring down the elephant.

This gets back to gambling and intelligent risk taking. I like to remind my client groups that in 1970, when I graduated from Colgate University, one of the dominant trends being discussed in the media was the perceived result of automation. The prevailing opinion was that mine would be the leisure generation. Machinery, like computers and robots, would make menial labor obsolete. The workweek would be four days (or less), and we wouldn't know where to bury the time. A key growth industry, therefore, would be the leisure business.

Now, unless Amy and I have missed something, the reality for our generation has been somewhat different from that projection. People now work harder, longer hours, with less job security than any generation in recent history. Nevertheless, the leisure industry has, in fact, been high growth because our leisure time has become so precious that we spend lots of money on ourselves, sometimes doing exotic things when we have the time. So, while the reality has been different than forecasted, the forecast that the leisure industry would grow has come true.

Focus Group Fallacies

Focus groups are probably the first tool that comes to mind when the idea of talking to your consumers comes up. They have long been one of the most accepted formats for collecting qualitative input from consumers and prospects. Everyone knows the format: a third-party moderator poses a list of preplanned questions to a panel of consumers while representatives from your company watch from behind a one-way mirror or on a video monitor.

Viewed through the spectrum of today's marketing environment, in which consumers are increasingly savvy about research methods, this traditional methodology needs serious rethinking. In reality, focus group participants know the unseen audience is there and perform for it.

So, what happens? Conventional focus groups generally run for one and a half to two hours because it's hard to keep anyone's energy up—on either side of the mirror—for longer with this kind of format. During that time the moderator runs the group through a series of exploratory questions. The questioning is either open-ended or it leads to a review of concepts and prototypes of new ideas. The result is usually a series of love it/ hate it responses that the moderator uses to gauge reactions to your new ideas.

What this research protocol really does is build a Berlin Wall between you and your consumers and prospects. Thus, the people seeking information and opinions are separated from the people who have information and opinions.

This needless separation was originally created because research empiricism held that a blind was necessary so that a company's name, reputation, and brand imagery wouldn't taint consumer responses. This builds on a false assumption: that qualitative research is in any way projectable to the marketplace. Projectability is not why you should be conducting qualitative research. In fact, while the budget for these conversations usually resides in the market research department, I wish it wasn't even called research.

What happens after focus group sessions end is equally bad. When the consumers leave, the client group usually sits around for a while informally talking about what they saw. This critical conversation is, generally, completely unstructured. Weeks later, you receive a written report from that third-party moderator summarizing her or his findings.

Who's kidding whom? You know that you haven't got time to wait for that report no matter how quickly it is expedited. Usually, within a week, another semistructured conversation takes place back at the ranch. Here, critical information is passed in a very second-hand manner. It tends to be shared anecdotally, with all sharers putting their own personal spin based on their own going-in biases. Decisions are then made.

No wonder so many decision makers hate focus groups and tend to discount what they hear from them! Why are so many mistakes made in qualitatively assessing consumer input? Here are three reasons:

1. *Because no moderator, no matter how well oriented, can ever know your business as well as you do.* Therefore, moderators will never be able to probe consumer clues, as they happen, as well as you can. If you've ever been one of the people hiding behind that one-way mirror, you know that few things are more frustrating than hearing an interesting comment come from a participant and then watching helplessly while the moderator fails to pursue it. By the time you send in a note or speak into an earphone (to prompt the moderator to follow up), the moment has passed. What could have been the beginning of a breakthrough idea is lost because the design of the research kept you from participating in a way that could have enabled you to probe and learn from the consumer's comment.

2. *Because, too often, the real decision makers are not present at these consumer conversations.* This results in misinterpretations and lost insights. It also reinforces another age-old practice in many corporations of killing the messenger, which occurs when lower-level participants try to bring home information that conflicts with firmly held beliefs of higher-ups.

3. *Because the need to report findings numerically defeats the purpose of the qualitative conversation you want to have with your consumers and prospects.*

The result, and the ultimate irony, is that this allegedly inexpensive form of consumer exploration becomes frighteningly costly in the long run!

A Better Approach to Consumer Dialogues—Co-Labs™

Here's what we recommend to overcome the inherent problems of traditional focus groups:

- Put yourselves in the room with your consumers and have a real, facilitated dialogue with them for up to four (that's right . . . four) hours.
- Make sure real decision makers are in the room actively involved in that dialogue.
- Don't quantify anything.
- Make sure you properly extract key learnings from all your colleagues via a facilitated session in which you discuss what everybody's

insights mean. We refer to this as the so-what session because you're answering the question, "So what does what we just experienced mean?"

When you become an active participant in an interactive focus group, which we call Co-Labs (a play on the word *collaborate*), the session is a much more effective and more immediate learning tool. In fact, after having conducted hundreds of these consumer discussions with company representatives as active participants, I can tell you, without reservation, that being part of the dialogue with consumers *always* enhances the information you extract.

We like to gather about ten consumers in a room, plus three to four client participants and a facilitator. We generally run two sessions simultaneously. In this way, six to eight people from the client team can participate in the dialogue, a critical mass that is necessary if the brainstorming that will come later is to be productive.

Yes, an interactive focus group involves time and effort by key decision makers and implementers. But the results are worth it. For instance, after attending just one of our interactive sessions, a senior executive from a banking client said, "I learned more about our consumers in this four-hour session than I have in 16 years with the bank."

Here is just one example of the kind of learning that can occur when you actually converse directly with consumers. A housewares client wanted to explore the topic of bakeware with consumers, so naturally we began the conversation focused on the kitchen oven. But it quickly became clear that what the consumers really were eager to talk about was their outdoor grills! This opened up a whole new area of market opportunity to produce products for use in grilling for this traditional marketer of kitchen equipment.

Breaking the traditional focus group paradigm by moving out from behind the one-way mirror allows you to discover and invent with your consumers. Their participation and viewpoints are essential to your properly understanding their issues and recognizing opportunities and in developing innovative ideas that will have a greater chance of success.

The interactive process is also much faster. All parties work together in real time. The key to speed, however, is in how you extract learnings from your team members.

By holding an organized minidebriefing immediately after each individual session, plus a so-what session after all the consumer sessions are completed, the information gathered is recorded and prioritized, results are summarized, and an action plan is created immediately. No more waiting two to four weeks while the moderator produces a third-person report. In effect, your team writes its own final report that is immediately actionable.

The Importance of Hearing with Different Ears

A critical by-product of the client group collaborating on the learnings is that it forces the group to understand and appreciate how each person hears things differently. This is due in part to our tendency to listen with our functional ears. For instance, marketers hear the marketing messages more clearly and might miss the engineering implications and vice versa for the engineers. You need to capture the different messages everyone heard to gain the full value of your conversation with consumers.

Several years ago, I captained a series of these four-hour, interactive conversations for Rich Products Corporation, the family-owned frozen food company I mentioned in Chapter 3. We ran several paired sessions with a cross section of restaurant operators in four cities.

At the end of the first day, we gathered the eight client participants in one of the session rooms for the daily debriefing. But before I could get the debriefing started, the head of R&D challenged the need to do it. "We've been sitting together all day talking to the same type of people in two different rooms. Didn't we all hear the same things?"

That gave me the perfect opening. I allowed as how that might be true and encouraged him to share the first key learning. I said, "Why don't you give me the first headline. If you could tell the people back home the one obvious truth that you learned here today, what would it be?"

I no longer recall exactly the headline he gave. But I do remember (quite fondly, actually) that the next three offerings from the group were either modifications of or outright disagreements with his obvious truth. After collecting those three or four additions, I glanced over at the R&D head, who looked very surprised as he grudgingly agreed that maybe there was a purpose to what we were doing.

In the end, what pleased him and the rest of the group was the way they began to develop a common set of agreement and disagreement themes. These then helped my cofacilitator and me to modify our plans for the next round of conversations with restaurant operators.

Our R&D client's initial impatience with the debriefing process raises another important point. You may meet internal resistance from the very start when you first raise the notion of trying to get consumers to spend four hours with you. Many people will be skeptical as to whether this is possible. But I can assure you that you will find that consumers are amazingly willing to help out if they are engaged in a meaningful discussion.

Consumers Want to Be Heard

Take what happened to me in the project I mentioned in Chapter 7, which involved developing new ideas for merchandise to be sold in a theme park. The person in charge of recruiting consumers in the park was almost speechless when I told him the sessions would last four hours. He couldn't believe I wanted him to ask people who had just paid to get into the park to give up half their day to sit in a room and talk. He almost stopped the project dead in its tracks. But I finally struck a deal with him: If the morning sessions failed, we wouldn't conduct the afternoon sessions.

Well, at the end of the morning sessions, we literally couldn't get the participants to leave! They were so invigorated that they wanted to keep the conversation going. They were saying things like, "I can't believe you really wanted to know what I had to say, but you did!" and "This has been one of the most exciting things I've ever done in the park." Needless to say, we conducted the afternoon sessions with the full support of the recruiter!

When to Dialogue with Consumers

You should begin your consumer conversations in the Discovery phase, to gather input in advance of the formal brainstorming work. But you will also be continuing these dialogues throughout the innovation process, as I will point out in later chapters.

After the Discovery dialogues, the key factor is in what you do with the information you gather. Rather than attempting to quantify what you've learned, try instead to listen *beyond* what you've heard for the underlying insights and "therefores" that should inspire your innovation efforts as you ideate for the future. Here are ways to use this information:

- Immediately after each four-hour session, and then again within 24 hours of the end of the last session, develop a list of your top ten learnings—the things you heard that caused you to say "aha" or to say "therefore, we can conclude . . ." or "therefore, we should explore . . ." Have fun with them. Use headlines and humor. Be provocative like David Letterman's top ten lists.

- Imagine being constrained to tell the members of the team who aren't present (including senior management) only the ten most important things you learned from these consumers. Each member of the participating team should do this independently with no input or influence from others. Then share your lists, discuss, and agree on a final list that will be your foundation of consumer insights for the brainstorming work.

Another powerful point at which to dialogue with consumers is after the brainstorming, when your beginning ideas are somewhat formed, but prior to starting development work in earnest. Be aware that there are helpful and not-so-helpful ways to do this. Beginning ideas are extremely fragile so you must proceed in a way that recognizes their vulnerability.

The least helpful way to discuss your beginning ideas with consumers is to review them in a traditional, qualitative research manner. Such an approach tends to extract bipolar, love it/hate it responses, thereby missing a huge opportunity to develop the ideas with consumer help.

Instead, use an interactive forum to invite consumers to help you enrich and improve upon the beginning ideas, regardless of whether they love or hate them. Here's how:

- Start by letting consumers know that loving or hating an idea isn't going to help you make it real or communicate it to other consumers. You need them to articulate the specifics of what they love and to build on the idea to improve it.
- Be sure to begin the evaluation with the positive side of the review; this also gives any idea momentum to help weather the negatives.
- Then, flip to the negatives. But, again, avoid the tendency to let the consumers complain about what's wrong with the ideas. Instead, ask them to discuss their negatives in ways that invite creative problem solving to make them better. For example, if you are reviewing an idea that involves unique packaging, their complaint might be, "I could never store a package like that where I store those things." Instead of concluding that the package design won't work, ask them what changes might make it more appealing or helpful or functional. Listen to their ideas, but, as discussed above, be sure to listen beyond

what they say for the deeper insights from which your knowledge base might benefit in modifying the package design.

Further down the line, when your ideas are more developed and you actually have a prototype to show, you may want to get together with consumers again. The danger here is in being so vested in the ideas you're presenting that you don't really listen to the feedback.

Here is just one example of a breakthrough that was made at a seemingly late stage of innovation because a company was willing to truly hear—and respond to—what someone in its marketplace was saying.

A few years ago, the people at Mr. Coffee invented an iced coffee maker for consumers. Like the earlier story about Vlasic's sandwich-sliced pickles, this one was a marketing no-brainer, but highly challenging technologically. Because cooling had to occur without diluting the liquid, it was far more complex than the drip method of making hot coffee.

When a working prototype was finally created, the product developers at Mr. Coffee decided to test the concept in some qualitative market research. During one interactive discussion, a retail store buyer suggested that the technology might be even more popular for iced tea, because consumers were more familiar with iced tea across the country than iced coffee, which was consumed mostly in the northeast (at that time). The result: an iced tea maker that was the most successful small appliance launched that year!

The people from Mr. Coffee did something very important here. As a result of this suggestion, they went back and expanded the definition of their business, beyond just coffee-related products, to include another beverage (tea). They might have ignored this suggestion because the idea conflicted with their perception that they were in the coffee-preparation business (comfort zone), instead of the beverage-preparation business (uncomfortable zone). Without going out and dialoguing with the supply chain, Mr. Coffee may very well have underexploited this opportunity. But they persevered, got closer yet to their consumers, and developed a truly innovative product in a segment that expanded their business base.

Market Immersion—Visiting the Real World

Management by walking around has long been considered an effective practice for executives to follow. Getting out onto the plant floor or visiting far-flung branches can reveal a world of information that would otherwise be missed by leaders who never leave their executive offices.

The same goes for what could be called innovation by walking around. Only in this case, the places you do your walking are where your product or service comes together with its end users—both at the point of purchase and in actual use. Physically immersing yourself in your market achieves several things that can have a tremendously positive impact on your innovative insights:

- When you step outside your work environment and start looking at the world that your consumers—and your products/services—actually inhabit, you get much closer to understanding your market than you can by almost any other method. Instead of looking at this world through the eyes of a marketer, an engineer, or a product manager, for instance, you begin to look at it through the eyes of a consumer or the eyes of someone who influences your consumer. This helps dissolve preconceived—and often strongly held—beliefs you might have that interfere with your ability to open up to new ideas.
- Competitive arrogance is a huge roadblock to innovation and one I will discuss in Chapter 11. Seeing no real competitive threats that could stop you dead in your tracks is not realistic thinking. Nothing deflates such conceit more effectively than being presented with evidence that you can't deny because you've seen it with your own eyes. This is what happens when you immerse yourself in the marketplace by talking to people you haven't really talked to before and observing things you haven't really observed before.

Be Ready for Surprises

Here's a quick example of the impact market immersion can have. A financial services giant wanted to innovate in the area of long-term health care insurance (LTHC). Our market immersion program for this client included visits to assisted-living facilities, nursing homes, and a long list of other places where the client group could talk with people who are involved in either delivering or recommending long-term care.

One of the people this group talked to was an estate attorney, because one of the first places many consumers turn to for advice about issues like long-term care is their lawyer. Our client is one of the bigger players in

LTHC, and was confident estate lawyers not only knew about LTHC but probably also knew about their company.

Imagine the group's surprise when the estate attorney said that LTHC simply wasn't on her radar screen! It was not something she knew much about and she had other solutions that she talked about with her clients. This discussion uncovered a huge gap between what our client felt attorneys knew and what they actually do know about LTHC. This raised the whole issue of needing to influence the influencer, something that neither the client nor anyone else in the industry had been actively doing.

Here's an important point to understand. Because our client's participants heard this information directly from that attorney, they could not ignore it as they might have if we had done a bunch of interviews and come back to them with a report about what people said about them and their product. That is what makes market immersion so valuable. If you are going to be honest with yourself, you have to take into account what you've seen and heard.

Now, let's face it. Not everything you're going to learn when you make a real-world visit will be positive. And, in some companies, the competitive arrogance is strong enough that people may try to bury any bad news. If you suspect your company is such an organization, having an objective, third party working with you on your market immersion exercises is essential. This person's role is to keep anything you learn that will be valuable for your innovation effort from being swept under the rug.

There are innumerable ways to observe your market and, yes, it is a lot of hard work. The main point here is that, no matter how you do it, you will always come back with an insight—probably lots of them—that will help drive innovation in the right direction.

For example, during a project to reinvent panty hose, three cross-functional members of our client's core team traveled through Europe, walking through hosiery retail outlets of all types, talking to consumers and retail store personnel.

Everyone was quite surprised when they returned having discovered a rather significant niche of male buyers of hosiery. And, while some of them were sexually motivated, a surprising number revealed that their doctors were recommending panty hose for athletic purposes and back/leg support. You would be amazed at how many Wall Street bankers are walking around with pantyhose on under their suit pants.

While this project was underway, my colleague, Jay Terwilliger, and I were also training some facilitators for a different client in Southampton,

England. During a friendly dinner at a wonderful old English pub, I mentioned to the people at our table that we were embarking on an exciting new project in panty hose and asked the group about English panty hose.

Jay and I were amazed at how much more knowledgeable the English women were about their panty hose than their American counterparts. For example, the hosiery industry buzzword for the thickness and weave of yarn used in making panty hose is denier. In our U.S. consumer work, virtually none of the women knew the term or what it meant. In Europe, almost every woman we spoke to knew what denier meant. In addition, they could describe the different feel and effect on their legs of different deniers, from elegant dress deniers to casual, everyday deniers.

We were also rather taken aback when Elizabeth, a rather staid member of the group, blushed and stammered before blurting out that she couldn't discuss it because panty hose was part of her erotica.

The point is that innovation by roaming around always unearths new insights into consumers and trade channels, their attitudes, and everyday practices.

By taking time to poke our heads out of our daily foxholes, by looking at and extrapolating from the seemingly unrelated trends we see, by engaging our consumers in meaningful dialogue, and by getting out into the marketplace and interacting with it, we can begin to develop educated hypotheses about where our marketplace is headed so that, like Wayne Gretzky, we can be there when the puck arrives.

In Chapter 11, I talk about the other stimulators of innovation: technology, competition, and government regulation.

[i n n o v a t i o n   f u e l]

- Make the satisfaction of your consumer's unarticulated desires your innovation goal. Realize that this is a much broader category to pursue than the oft-referred-to category of unmet needs, of which most Americans really have very few nowadays.

- Seek qualitative not quantitative consumer input during Discovery.

- Determine how you can lead rather than follow consumers by:

  - Exploring trends, including broad-ranging ones that might not at first appear to have relevance to your market.

  - Replacing focus groups with Co-Labs, or interactive alternatives that remove intermediaries from the conversation between you and your consumers (past, current, and potential). Use these facilitated dialogues to uncover unarticulated desires and wants.

  - Spending time in the places where people buy and use your products/services.

- Honor the fact that consumers want to be heard; talk with them and then delve behind their words for the hidden insights that can inspire breakthrough innovation.

CHAPTER ELEVEN

Discovery (Part 2)— Competition, Technology, and Regulation

Competition

In what markets do you compete? That probably seems like a simple question but, believe me, plenty of companies don't answer it correctly. Many define their markets one way while consumers define them another way. Because consumers are the ones with the bucks to spend, their viewpoints on what markets a company competes in (i.e., what purpose the company serves in the consumer's life) should be considered the final word. But too often, it isn't. This is part of the innovation roadblock I call Competitive Myopia.

Here's a perfect example of Competitive Myopia. Years ago, when ramen noodles first hit the market, one of the leading market research firms that tracks market share went to the major soup manufacturers and told them they had a problem. Ramen noodles had really taken off and the ramen brands were beginning to erode competitive market shares in the soup category. In a grand demonstration of competitive myopia and bravado, the head of one of the major soup companies at the time declared that ramen noodle products were not soup and instructed the research company to redefine the soup category so that noodle products like ramen weren't included.

The good news: The company was able to maintain its market share in the soup category. The bad news: While its share remained constant, sales volume declined. Consumers didn't care if ramen noodles were not soup according to the big soup producers. Consumers liked them and they met consumers' needs for a quick, easy-to-prepare mealtime alternative to soup.

That soup company could have taken a different approach when the research firm pointed out that it was losing market share to this upstart alternative. It could have chosen to compete in the ramen market and, given its marketing power, might easily have driven out the less-established ramen brands. But fighting it out with new competitors is not a risk some companies want to take. Too many prefer to arbitrarily narrow their market definition in ways that the consumer could never imagine to the point where they can proudly proclaim they are the dominant leader in that market. The inescapable truth is that the consumer always decides.

Killing Innovation with a Narrow Market Focus

The desire to be the biggest fish in your marketing pond, even if that means you have to arbitrarily drain water (and would-be competitors) out of it, pervades many industries. Achieving market dominance by very narrowly defining who it is you are serving and what it is you are providing shuts off possibilities to meet new needs and capture share in new markets. It can kill innovation by:

- Closing your ears to the voice of consumers who may have new wishes and desires that you could fulfill if only you weren't so focused on maintaining market dominance above all else.
- Narrowing your innovation focus so much that you ignore ideas that might lead to breakthroughs in related markets.
- Blinding you to competitive threats from companies who define the market differently than you do—perhaps more closely to the way consumers define it.

Look, for instance, at what happened when the fax machine came into widespread use. At the time, FedEx, Airborne Express, DHL, and the U.S. Postal Service failed to anticipate the impact the fax would have on their companies. They were busy competing with each other for market share of the overnight delivery business while consumers were open to new solutions from completely outside of that market. FedEx did wake up at some point and tried to jump into the fax market by setting up faxing centers. But by the time this strategy was put in place, the price of fax machines had dropped so much that every business (and now homes) could afford to have one in-house.

How Do I Spend This Dollar?

Now, obviously, I'm not saying you shouldn't pay attention to market share. Being a market leader does have big benefits. What I am suggesting is that market leadership is not valuable if the only way you can achieve it is by maintaining a very narrow market definition. You may be leaving a lot of money on the table to be scooped up by companies that are more focused on how to better serve consumer wants and desires. And you put yourself in danger of having your market taken right out from under you by new competitors who find a better way to serve your consumers while you focus your energy on maintaining and promoting market leadership in your market niche as you choose to define it.

Remember this: When presented with a great new choice in the marketplace, no consumer thinks, "Gee, I should probably stick with the market leader even though this new product would serve my needs better."

In the end, consumers have the final say. They decide how to spend their dollars and they don't particularly care how you define your market or who you consider your competition to be. If someone comes along from outside the market as you've defined it that will serve their needs better than you, you're history.

Some companies understand this very well. In the mid-1980s, Hallmark Cards shook up the advertising industry when it suddenly fired its advertising agency, Young & Rubicam, within a week of the agency signing AT&T's long-distance unit as a client. Hallmark's stated reason was that it didn't believe one ad agency could serve two companies that use the same types of emotional, motivation ads.

At the time, many people in the ad business scoffed at Hallmark's rationale. But I believe those scoffers failed to see the very real conflict of interest that existed between the two agency clients. While many view Hallmark as being in the greeting card business, the company had the intelligence to realize it is also in the reach-out-and-touch-somebody business. Consumers can just as easily meet this need by making a long-distance phone call as they can by sending a greeting card.

Curing Competitive Myopia

Unfortunately, Hallmark is the exception rather than the rule in terms of companies that understand the importance of defining their market from the consumer's point of view. Adding to the problem, many companies don't even have any really clear idea of what their competitors are doing within the narrow markets they have defined for themselves. A surprisingly large number of companies don't bother to make competitive intelligence a priority. In fact, for too many companies, competitive intelligence is an oxymoron. Because they are blind to threats from within their self-defined markets and oblivious to threats from outside as well, such companies run the risk of getting blindsided in the marketplace.

One focus of your work in Discovery must be to overcome this tendency toward Competitive Myopia. Chapter 10 showed you ways to see your market through the eyes of the consumer so that you can define your market in new ways that will spur innovation. Equally important to your ultimate innovation success is that you must learn as much about your existing and potential competitors as possible.

There are a number of ways to do this that address what I call the logic-instinct spectrum. This spectrum suggests that our ability to process information falls along a spectrum from pure logic and fact-based reasoning on one end to total gut instinct on the other end. In Discovery, the goal is to conduct as many exploratory exercises as possible that will probe both of these informational extremes. I described several in Chapter 10 for exploring consumers. Here are some new ways to investigate your competition with an open-minded inquisitiveness that you may not have used before.

Competitive Information on the Internet

Fortunately, the Internet has made it easier than ever before to gather competitive information. Any innovation effort should include a thorough search for information about competitors that can be used to spur your brainstorming. I am increasingly astounded at the richness of competitive information sitting right there on the Web.

Begin by searching for known competitors, by name. Look at their Web sites and search news sites for articles featuring their names. Don't forget to look at Web sites such as BusinessWire.com, where you can find press releases posted by your competition. This will address the logical end of the spectrum in your Internet search.

But don't stop there. For example, let's imagine that you *think* you are in the ergonomic office chairs business. A search engine can help you identify manufacturers you don't know about including ones in other parts of the world.

You will discover all kinds of information on the competitors that you have listed and the new ones you discover in this search. And don't just focus on product information. Look at the services other companies provide and at how they are distributing their products. Both of these areas are ripe for innovation.

Your process might look like this:

Searches to be made:

- Ergonomic office chairs
- Known competitors
- Unknown competitors identified by search engines

But keep going! Now use the Internet to stimulate your gut instinct. Consider running several additional searches based on key words that your customer might use to describe what you do for them. You should be very surprised at what you will find. Your list might look like this:

Customer words related to my business:

- Chronic back pain
- Arthritis
- Home office furniture
- Back care products
- Snack and drink holders
- Blankets and comforters

Enter these words into an Internet search engine. Don't be afraid to follow the word trail wherever it might lead you. You may feel a bit like Alice in Wonderland following a rabbit into a whole new world of discovery.

For example, when we entered back care products into our Internet search engine, it revealed a company that makes travel accessories for people with back problems. Because you're already selling to businesses, making products for business travelers might be a good extension. Connections like this might spark an avenue for new opportunities for your ergonomic chairs business. Who can say? What's important here is that the

Internet offers all sorts of opportunities for specific information gathering (like seeing what products and services your known competitors offer) and for just random surfing that leads to connection making that can provide lots of grist for your idea mill.

Competitive Mapping

Another way to balance logic with instinct is to find the so-called white spaces in your competitive geography by the creation of a series of competitive maps. The white spaces—the areas that have yet to be filled—on those maps, if created properly, will help your team identify and exploit competitive opportunities.

Each map takes any two of your industry's key differentiators and plots them as coordinates on X and Y axes to show where you and your competitors stand in terms of strengths and weaknesses as displayed in Figure 11.1.

The entire mapping exercise is driven by the parameters chosen, which makes it very vulnerable to a Garbage In/Garbage Out result if company biases and myths drive the exercise instead of rigorous, sometimes painful honesty. The parameters should be those elements that are the industry's

[FIGURE 11.1] Competitive Mapping

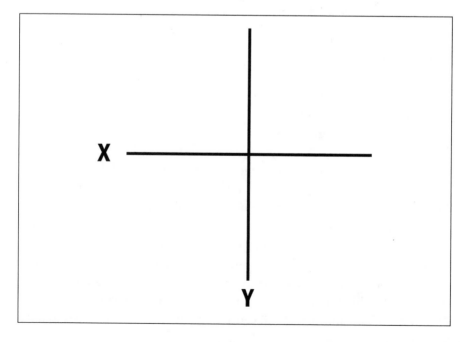

most important differentiators. The X and Y parameters should contrast one another, and each should be able to have a range of values. Some examples: speed to market (slow to fast), convenience (hard to easy), customization (personalized to mass market), flexibility (inflexible to highly flexible), cost (low to high), and variety (none to lots).

You will know if you have chosen the right parameters for your maps when everyone in your competitive sphere is not bunched up in the same quadrant. If this happens, then you have not chosen competitively meaningful parameters. Keep experimenting until the maps start to tell you something.

This look at your competitive landscape will allow you to assess the potential impact on your innovation effort by the interaction of different criteria. By locating where the white spaces of opportunity are, you will help ensure that your brainstorming in the Invention phase is directed toward areas with real potential. While you cannot completely rule out the threat of competition, particularly in a business era where any market edge seems to last only a nanosecond, you will at least be headed toward sectors that play to your competitive strengths.

You can have any number of these maps depending on how many factors you define as being critical to your success. As you work to create a vision of your competitive landscape through these maps, be sure to consider factors related to each of the other drivers of innovation—consumers, technology, and, if pertinent to your industry, government regulation.

You may meet resistance to focusing on what consumers value. Engineers, for example, may tend to want to define what is important based solely on the capabilities of their technology rather than on what consumers really care about. But if you focus your thinking that narrowly, you will not uncover white spaces representing real opportunity in areas where technology solutions are not the answer.

As you will have gathered from what I said earlier in this chapter, it's critical to look beyond the players in your present market. Consider companies from outside your market that might present a competitive threat because they serve similar consumer needs. As suggested in the earlier story about overnight delivery services and the fax machine, be on the lookout for your equivalent of the fax, an unrecognized threat from outside your field of usual competitors that could quickly eat big holes in your market.

One last point about mapping. Dwight Eisenhower once said: "The plan means nothing. Planning is everything." This is especially true for any mapping exercise. It's not about whether the maps you create are right or wrong, it's the richness of insight gained by the group from the conversations around each map and the so-what thinking that is driven by the exercise.

Try Roaming Around

As discussed in greater detail in Chapter 10, nothing exposes Competitive Myopia as vividly as field trips into the marketplace, and I am constantly surprised at how rarely this happens. The simple act of observing consumers where they purchase and use your products and services and those of your competitors can be powerfully eye-opening. One of our client's core team members calls these marketplace visits "a day in the life." We often attach such a day to other Discovery work and, like Co-Labs, conduct a so-what session to mine learnings from the immersion experience.

Narrow-Minded about Technology

Technology is another major driver of innovation that you must study in Discovery. As with competition, organizations almost always define their technologies too narrowly. They tend to think of them only in the context of hard technologies, such as manufacturing or R&D capabilities. Too often, they don't talk about technologies that are related to providing service and other forms of consumer satisfaction despite the fact that such capabilities are increasingly becoming the only true, long-term differentiators in the minds of consumers.

It usually starts at home with a silo mentality that prevents us from exploiting technologies that exist right within our organizational framework. Big companies, with multiple divisions, rarely consider how they can migrate technologies from one business unit or sector to another. Most often, people in one division aren't even aware of what technologies are available in other divisions. In effect, they make the same mistake that small companies offering only one product line make. They act as if they have only one technology—the one that produces their one service or product line.

This narrow-minded approach to technology is another kiss of death when it comes to innovation. When thinking about innovation, you need to consider all of the technologies that could help you deliver your product or service, including technologies that might be available elsewhere in your organization that could be put to work solving your innovation problem. And, too often, they are the ones that initially appear to be irrelevant.

Circles of Technology

In Discovery, you must explore technologies as if they fit into three concentric circles as shown in Figure 11.2.

In the first circle are the technologies that you have in your own back pocket, the stuff you currently use to produce and deliver your products and services. In assessing these capabilities, be sure to identify all of the tools and techniques your organization possesses for purchasing, produc-

[FIGURE 11.2] The Circles of Technology

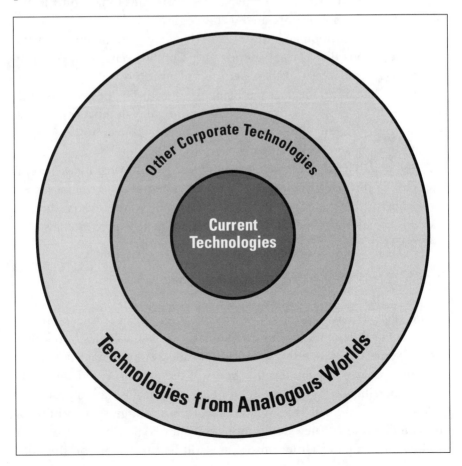

ing, selling, billing, distributing, repairing, and servicing what you do (depending on your industry, I may have left out one or two steps). Don't leave out anything for which you have a capability or a system. Expand your definition of technologies. Remember that no company has only one technology. If you think you do, you probably won't survive much longer.

For example, over the years your organization may have developed raw materials purchasing and manufacturing capabilities that enable you to compete as the low-cost producer in your business. By my definition, this could be something on which to build competitive capabilities in whole new arenas.

The second circle is where you should explore all of the technologies that are used in other parts of your organization. This can get very exciting when your business is part of a multifaceted corporate entity. The challenge comes in opening up your mind (and those of people in other parts of the corporation that have the technology you want to tap into) to the value of these seemingly unrelated technologies.

For example, a popular industry story has it that someone from the Mars snack foods division got the idea for Combos (cheese-filled, mini-pretzel logs) from a visit to their pet food plant where a line of marrow-filled, bone-like treats for dogs was being produced. The potential for this kind of technological crossover is available every day in most companies. We simply miss the invisible obvious by not opening our minds to the potential for using something in a new way.

In the third circle is technology from analogous worlds. I gave several examples of this back in Chapter 5 when I talked about bringing in experts from outside your company and even outside your industry who can share information that might lead to new insights about technology.

Tapping the Three Circles of Technology

Your objective in looking at this mix of technologies is to explore unusual ways of linking all of your capabilities with technologies from other parts of your corporate entity and from outside as well. What will emerge is an array of new capabilities that you will then use to stimulate breakthrough possibilities in the Invention phase of your innovation process.

Some potential capabilities will be easily within your grasp, consisting of new combinations or new applications of technologies that already exist within your company. Other possibilities will be further out on the horizon. These might include technologies that are used in other industries that will require some work to adapt to your field or may be technologies that haven't been perfected yet.

A key to remember here is that, at this point, you are not concerned with whether the technological possibilities you are exploring are marketable or could be used profitably. Your objective is to come up with new capabilities that the world is not seeing right now but which might possibly be available somewhere down the road. This refers to that old Bell Labs mentality I discussed in Chapter 9.

Try It; You'll Like It!

Don't be surprised if you have a hard time getting people to agree when you say you want to assemble a group of technology experts from parts of your company who have never come together before. Clients routinely are reluctant to agree when I first suggest that they might benefit from this cross-fertilization of technologies. Yet, when they hear a few examples of the kinds of breakthroughs that can occur when people from disparate technology backgrounds come together, they usually agree to give it a try. And, here's the important part: I have never had a client say later that they didn't receive real value out of including as diverse a range of Wild Card experts as possible in the Discovery phase technology sessions.

The Regulatory Piece of the Innovation Pie

The fourth factor that stimulates innovation is the regulatory and legal environment. Obviously, this applies more strongly in some industries than in others. But no matter what your industry, government policies can either prompt a need for innovation or negatively impact your ability to innovate at all. And clearly, if you're in a heavily regulated field, you need to consider the regulatory implications as you move forward with innovation.

One of the main problems that arises when companies start to consider this topic is the us-versus-them mentality that is so prevalent when most people think of regulators and regulations. Often, this adversarial kind of thinking even extends to the lawyers within our own companies because they are often the ones who bring the bad news about how regulatory issues will impact a proposed innovation. Thanks to the kill-the-messenger attitude that many of us hold, the lawyers are viewed as people who are more likely to obstruct innovation than to support it.

The reality is that including lawyers and regulators in the Discovery and the Invention phases can have real benefits. Although I often have to fight to make this happen, I value having lawyers included on the innovation team. They're usually bright individuals who have been trained to look at the world differently and who are comfortable putting forth ideas. These are just the kind of people you want to have on your team.

By including the lawyers as part of your team early on, you can avoid many problems that invariably arise down the road. Rather than waiting until that time when their negative reaction to an idea in which you've invested a lot of time and money can be destructive, you will enable your legal team to proactively problem solve around regulatory and legal issues in the early stages of idea development. And believe me, they will appreciate not having to play their usual role—the guys in the black hats who kill ideas at the last minute.

I have found that often lawyers can be your strongest friends when you find yourself in the Dark Night of the Innovator. They know how to advocate for ideas so that once they're committed to something, they can be very effective at helping you push it ahead through the maze of corporate obstructionists.

It is also sometimes possible to include regulators in your Discovery work. Not long ago, I worked with a client whose business is closely tied to a large government agency. Without the agreement of this agency, the client's ability to innovate in some business areas was limited. So we invited a few people from the agency to participate in brainstorming. They not only provided great insights but they also were extremely excited by the way the innovative direction in which the client was moving would help the consumers served by the agency. So instead of being obstructionists, they started working on regulations that would support the innovation.

By working in a collaborative fashion with regulators whenever feasible, you may shave time off of the Implementation phase of your work because regulatory hurdles could be cleared from your path at earlier stages. Obviously, such an approach is not possible in all cases. Some regulators may not be open to the idea for fear of being accused of supporting one company's interests over another. But at least keep your mind open to the possibility that regulators may be not have to be the enemy in all cases.

[innovation fuel]

- Don't fall into the trap of artificially defining your market segments too narrowly because being too narrowly focused might lead you to ignore potentially fruitful new areas of exploration.

- Find out how consumers define your market. You may learn that you are competing against people you never considered before.

- Fight Competitive Myopia by getting out into the marketplace, and by using Internet keyword search engines creatively and following the word trail wherever it might lead you.

- Also use competitive mapping to avoid Competitive Myopia. Look for the white spaces on these maps that indicate areas where your competitors are not strong and are thus areas that are ripe for innovation.

- Expand your perceptions about the technologies that are available throughout your own organization and ones that will be useable in the near future. Also, explore technologies from other fields that might have potential in yours. By looking at the three Circles of Technology, you may discover new areas for innovation that weren't apparent to you before.

- Avoid the us-versus-them attitude when thinking about regulators or about your own legal team. Consider whether it might be possible to have regulators participate in the Discovery and Invention phases of your innovation effort and make use of the special skills that lawyers can bring to the innovation table.

[PHASE THREE]

Invention

CHAPTER TWELVE

Invention (Part 1)— Divergence

H aving completed the due diligence of Discovery, your innovation team members' minds are now well primed for the all-important third phase of innovation—Invention. Grounding Invention in the learnings and insights gained during Discovery, in a way that promotes and stimulates speculative thinking, is one of the critical success factors for Invention. When organizations exploit the power of Discovery during Invention, ideation becomes an extremely positive and productive experience for everyone.

Besides building on your Discovery output, the other critical success factors for the Invention phase are:

- Using championed teamwork as the decision-making model for a well-chosen, cross-functional team of champions, creators, and doers
- Having a well-defined yet unmeasurable core purpose and core values for the business *and* a measurable and motivating task focus/BHAG for the innovation initiative, as touchstones for decision making
- Tapping the power of the Overnight Effect
- Selecting potentially big ideas based solely on the criteria of potential for breakthrough newness at the point of wish selection
- Providing for the evolution of an absurd idea into a potentially breakthrough idea via a structured technique for open-minded evaluation

I already introduced the importance of championed teamwork as your decision-making model for innovation in Chapter 5. By this time, having worked together through the Discovery phase, your core innovation team should now be comfortable and confident with how championed teamwork supports and will ultimately increase its chances for success.

Chapter 7 also explained why, in the absence of numbers, a strategic core purpose, a set of unchangeable core values, and a clear task focus for any innovation initiative are critical to your success. Without them, you will be rudderless.

Although it's not on my list of key success factors, one of the most important parts of Invention is that everyone has a rollicking good time. My coauthor, Jeanne Yocum, loves one of my many politically incorrect sayings, which is that "Invention is the most fun you can have with your clothes on." Having participated as a Wild Card in several of our client sessions, she knows whereof she speaks.

A Cycle, Not a Session

Now, let's explore Invention to help you understand where, why, and how the other three critical success factors matter. An important distinction to make at the start is that Invention never occurs in just one meeting but rather in a series of sessions, which I think of as a cycle. This addresses the mistaken (and terribly frustrating) belief that it is possible to invent breakthrough ideas in one meeting or session.

Again, I am constantly astounded at the number of the people who ask us to help them reinvent a business, a category, or a brand in one meeting. They often say something like: "We'll set aside an *entire* day for this." Their tone implies that a day should be more than enough time and any expectation that they spend more time away from their regular work would be unrealistic. But what could possibly be more important than inventing a company's, a division's, or a brand's future? It takes time!

In Discovery, you take the time to fill your brains with information about the four drivers of innovation (consumers, technology, competition, and regulation). Instead of a few flurries of thinking, your brainstorming now has the potential to generate a blizzard of ideas focused on the task. When this happens, you will need to be prepared to deal with more ideas, newer ideas, and just all-around better ideas than if you had jumped into ideation prematurely without taking time to add this important fuel for ideation. So, how can you expect to process all of this rich information and use it to drive breakthrough, new ideas in one gathering? It can't be done!

If you allow too little time, this short-sightedness will have the same effect as someone who drives a sports car while, at the same time, putting a firm foot on the brake pedal and pushing the accelerator to the floor.

The Overnight Effect

One important reason for not trying to do all of your Invention in one day is that you'll miss out on the benefit of one of the critical success factors—the Overnight Effect. This is a simple, yet powerful, psychological phenomenon that dramatically improves the output of any invention effort.

The ability of your team to generate great ideas—representing real potential breakthroughs—will grow exponentially if you build at least one unstructured overnight into every session so that every meeting flows over two or more days. (This advice applies to not just Invention sessions but also Discovery sessions.)

During this overnight period, people's minds always operate in the relaxed concentration mode described in Chapter 1. Bits of information come together and new mental connections and concepts are formed that wouldn't have time to form if ideation occurred in just one day. I often go so far as to push a client to begin a one-day segment of Discovery or Invention on an afternoon and finish it the next morning, so that they can always benefit from the Overnight Effect.

All of us have experienced the Overnight Effect, usually without realizing it. We've gone to bed thinking of a problem and, presto, in the shower the next morning a great solution becomes apparent. Many people say that they get their best ideas in the shower, thinking that somehow the shower is the idea stimulus. But what they're really experiencing is the Overnight Effect. The output of new insights actually developed during the night yet that only became conscious during the first morning activity—like the shower.

We always start the second day of any session by asking team members for their overnight thoughts. Time and again, I've seen that the best ideas to come out of ideation efforts are the ones that people present at this point. Do not shortchange your invention potential by designing a time frame that doesn't allow for the Overnight Effect.

The Two Stages of the Invention Phase

The best design for ideation honors what the mind instinctively wants to do, which is to start by getting all the possible ideas out on the table and then narrowing them down to a few, high-potential possibilities. This pattern of divergent followed by convergent thinking makes up the two stages of the Invention phase. These two stages allow for

1. uncensored explosion of divergent thinking followed by
2. carefully controlled convergence and idea development, refinement, and selection that leads to breakthrough innovation.

In essence, this divergence/convergence process looks like a diamond (see Figure 12.1). You start (at the top) with your task, followed by a bit of background information. You then encourage the group to generate as large and uncensored an array of wishes (using the phase "I wish . . .") as possible, building on each others' ideas until you have a wide selection to choose from. Then you carefully select from all those ideas only the few

[FIGURE 12.1] The Six Steps of Invention

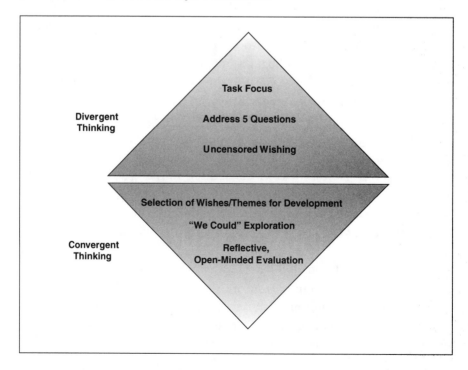

that you want to develop further. You then problem solve to develop those wishes into doable ideas.

While this approach has historically been used for creating new product and service ideas, it also is fully applicable for all other forms of innovation, including the setting of strategic direction, the exploration of manufacturing process improvements, cost reduction, service improvement, or, in fact, any part of your business that you wish to improve.

By the time you've completed a successful Discovery phase, team members should be on the edge of their seats, ready to let loose with the ideas that have been bubbling to the surface of their minds as a result of everything they have experienced. The Divergence stage is where you introduce the brainstorming process you will use to capture this avalanche of ideas.

Divergence—Letting Ideas Flow

The first stage in the Invention phase, Divergence, is obviously a very important part of the innovation process. It is where you will begin to connect all of the output from the Discovery sessions (see Figure 12.2).

Generating and developing exciting, wild, and crazy new ideas together is a mentally stimulating, freeing activity that energizes, empowers, and unites any innovation team. The momentum gained during Divergence can go a long way toward building up the head of steam that is needed to push ideas through the Dark Night of the Innovator.

[FIGURE 12.2] Divergent Thinking

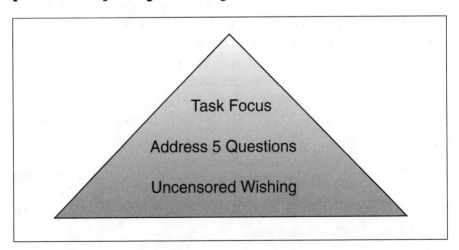

Task Focus

Address 5 Questions

Uncensored Wishing

To have as much creativity in the room as possible for the Divergence stage, the core team should be supplemented with other people from a variety of functions within the company, as well as some Wild Cards (as discussed in Chapter 5). You will remember that Wild Cards are those unrestrained outsiders from other parts of your corporate entity and outside the company who look at the world differently and can help any group spin intriguing mental pictures of the future. These people don't draw a paycheck from you, and because their careers don't depend on your approval, they feel free to be unorthodox and uninhibited. Wild Cards always stimulate everyone's thinking and they help get the ideation process off and running at a faster pace.

It is also wise to include several people in this session that you found to be particularly instrumental in any of your Discovery sessions. For example, a creative technologist from a technology exploration session, a particularly observant market researcher from your marketplace travels, a very open-minded trade channel customer that you found during a field trip, or an inventive plant manager. These people will help you bring all of the thinking that occurred in those Discovery experiences into the Divergence stage beyond just the nuggets. You will use the nuggets to get the information ball rolling but you can count on Discovery parti-cipants to add much more information and ideas along the way during Divergence.

Starting Divergence—The Five Questions

It is very important that you start the divergence/convergence process for ideation by presenting your innovation task focus/BHAG to the ideation team and addressing only five basic questions. Core team members will be thoroughly familiar with this task, of course, because they will have participated in developing it. But any new team members, especially Wild Cards who might be brought in only for the Divergence session, will need to be brought up to speed on the purpose of your innovation effort.

After presenting the task focus/BHAG to your invention group, take no more than 10 to 15 minutes to answer these (and *only* these) five questions:

1. *Why is this an important challenge to work on?* What will happen if you are able to take advantage of this opportunity or solve this problem? Also, what might happen if you fail to take advantage of the opportunity or solve the problem?

2. *What, historically, is important for the group to know?* Briefly, provide only the truly critical background information. Believe it or not, those present who did not participate in the Discovery phase do not need a half-hour dissertation on your topic. The nuggets that will be presented as part of Invention will be sufficient to get their minds rolling. So, fight the urge to do a total brain dump on them at this point in the process.

3. *How will decisions be made for this endeavor?* If a decision maker is not present for the ideation session, the team should be made aware of this before ideation starts. Team members need to know, in advance, if the ideas generated are going to undergo review and selection by absent leaders. In other words, they should be aware of just how empowered the group really is so they don't have false expectations for the outcome of their ideation work.

4. *What's already in your head?* Answering this question helps avoid having the group reinvent the wheel and gives you an opportunity to provide examples of the types of ideas you're looking for. If you want people to shoot for the stars in their ideating, this is your opportunity to model that behavior by giving some examples of potentially breakthrough yet flawed ideas that may be floating around in your head.

5. *What do you want from this meeting?* Here you define your hopes for the session (and the overall project). I like to coach my clients to avoid the extremes in answering this question. By that, I mean that they shouldn't say either, "I'm looking for the BIG idea," or, "I'm looking for 500 ideas." This tends to create performance anxiety in the group. Instead, I encourage them to paraphrase the slogan for the U.S. Marines by stating that they are looking for "a few good ideas."

Why Only These Five Questions?

Let me explain, with another story, why briefly answering only these five questions is so important. It's a classic example of the "difference between intent and effect" speed bump introduced in Chapter 8.

In 1973, I was a second-year MBA student at the University of Michigan Business School. As a marketing major, I was very excited about taking a course that promised a real-life business experience. Each semester, the professors who taught this very popular course arranged for a nationally recognized corporation to challenge the class with a real business problem.

The class would then break into work teams to create marketing solutions for the company, which were presented to that company's management team at the end of the semester.

One reason I was really looking forward to this course was that the prior semester's class had worked for Olin Skis and everyone in the class had experienced a free one-day demonstration ski trip. Imagine my disappointment when the announcement came that our client company would be Gerber Baby Food!

But, here's the reason for my story. The senior management of Gerber arrived on the first day of class armed with reams and reams of paper for each team. They spent two 90-minute class periods, plus a field trip to their headquarters in Fremont, Michigan, detailing their challenge. Yet, with all this rhetoric, they never presented us with one simple, clearly articulated invitation to innovation.

This was a shame because their challenge was really quite simple. Had they used our concept of a task focus (an invitation to innovation), the task focus could have been to develop creative marketing ideas to get non-babies to eat Gerber Baby Food. That was what all the talking boiled down to—an easily understandable 12-word challenge.

In addition to not providing a short, well-stated task focus, Gerber management compounded the problem by making it clear that they wouldn't entertain any suggestions for modifications in their current offerings; like changing the product (make it chunkier), enlarging the jars (for larger appetites), creating line extensions with graphics of older people, pricing variations, in-store placement alternatives . . . nothing!

We didn't need three hours of data download and a trip to the plant. We could have started after 15 minutes with a clearly articulated task focus and the answers to the five questions listed above. Gerber's intent was to motivate by giving us all the available information but the effect was to create the Information Overload roadblock that left our minds numb.

These five questions have been honed and refined by facilitators for generations. Their purpose is to extract enough information to stimulate ideation without overloading or constraining the team.

You don't have to wait for a big official innovation effort to try this method. Test it out in any meeting where you'll be brainstorming. Present your team with a simple task focus invitation to innovation, in 15 words or less, plus the answers to only those five questions. Then jump right into the wishing phase of the brainstorming. If questions arise, encourage people to express their confusion in a wish and not worry about being right.

Tell them to guess what the answers might be and use those guesses for more wishes. You'll be surprised at how many ideas you can get out on the table in a short period of time. And as you use this method again and again, people will become more skilled at wishing and the quantity and quality of ideas will increase even more.

Shoot for the Stars

Once the five questions have been answered, you can begin to use the nuggets from Discovery to promote speculative thinking. Core team members and other Discovery participants should be prepared to present the nuggets from the Discovery sessions in short bursts throughout the first day of Invention. The meeting facilitators should then invite the team members to wish for anything that pops into their heads that might be a way to address the task focus using the Discovery insights.

I cannot overemphasize how important it is to encourage people to shoot for the stars with their wishes. Here again is where your group work in Discovery should pay off because, by this time, team members should be thoroughly comfortable with each other and feel that the environment is safe for experimentation and even for outrageousness. (Remember the I-wish-for-a-bagged-whipped-cream-that-looks-like-a-zit story from Chapter 3?)

It can't hurt to remind people of Oscar Wilde's admonition that "an idea that is not dangerous is unworthy of being called an idea at all." Good facilitators will often encourage people to come up with ideas that are illegal, immoral, or would get them fired for being offered up outside of this room.

It also helps if the team's leaders model absurd thinking. In fact, nothing gets things rolling creatively like a wild and crazy, absurd wish from one of these people early in the session.

For example, several years ago I was working with a team from a company that makes peanut butter. The task for the project was to develop new products and line extensions for the brand. After some skills training, its leader, a somewhat buttoned-up young woman, decided that she wanted to really stretch her team's permission to think outside the box. So, when we began her wishing session, she demurely raised her hand (with a twinkle in her eye) and wished "that we could use [our peanut butter] as a sexual lubricant."

You can just imagine what happened in that room. The first reaction was about three seconds of dead silence. Next, the room exploded in laughter (the good kind). It actually took me almost ten minutes before I could restart the session. During that time there was a mix of humor, silly builds on her wish ("Think what we could do with Chunky!"), and other over-the-edge responses.

But guess what happened to that group? Their creativity took off. Their ideas were all over the place. Their willingness to say anything was unbounded.

Now, here's the point. When it came time to choose ideas for further development, did anyone pick her lubricant wish? No, of course not. But did that one single act by a risk-oriented team leader pay big dividends by giving her group permission to mentally stretch? You bet!

Ground rules for brainstorming, including the all-important "No Bazookas," should be thoroughly ingrained in people's behavior by now, and this should support strong ideation. Of course, old habits die hard so the facilitators should remind people of the ground rules at the start of Divergence. (See Appendix A for some of the ground rules we use at Creative Realities.)

Throughout Divergence, the facilitator must also be vigilant for innovation speed bumps caused by ground rule violations and be ready to quickly initiate corrective action. If you have created the right environment, invention team members will feel free to enforce ground rules themselves. When people spot bazookas and feel comfortable in pointing them out to the offenders, even if the bazooka wielder is the boss, you know you've got the right climate for ideation.

Remember, however, that if ground rules are not enforced, you will not get full value from your ideators because people will be unwilling to take chances and will become guarded about sharing truly novel ideas. Equally damaging, if team members feel the session was hampered by speed bumps and broken ground rules, they may not develop a strong sense of enthusiasm for the beginning ideas produced. Establishing this energy and the sense of ownership that accompanies it will be critical to your ability to drive the effort through the Dark Night of the Innovator.

Simulating What Great Thinkers Do

You may recall that the Rich Products story about the bagged whipped cream that looks like a zit resulted from an exercise I led that began with the description of favorite movie scenes. The movie scenes led to connection-making activity, which led to some absurd wishing.

That exercise is one example of the facilitator's most powerful creativity-stimulating, behavioral tools. My first partner at Creative Realities liked to call them managed serendipity. Others call them excursions, blitzes, brain farts, or side steps. I like to describe them, paradoxically, as controlled mental flights of fancy.

These exercises in absurdity have evolved over the years as a way of simulating the thought processes of great thinkers by stimulating focused minds with seemingly irrelevant stimuli that force the brain to deal with any problem or opportunity in a new way.

They can be very helpful at many stages of ideation to spur greater creativity. One of the most popular times to use them, of course, is the wishing phase during Invention. Here are the five basic steps for running this type of exercise:

1. Forget about whatever you are working on and put it out of your mind.
2. Use any creative device (like scenes from a favorite movie or an object in the room or a recent news story) to generate some seemingly irrelevant thoughts.
3. Play with those seemingly irrelevant thoughts by forcing your mind to make mental connections (linkages) between them and anything that pops into your mind.
4. Use those connections/linkages to drive an absurd idea or wish (illegal, immoral, you'd get fired . . .) on your task or topic.
5. Extract a principle or element from your absurd idea and use it to drive a second generation, closer-in idea or wish.

Consider rereading the Rich Products story once again (see Chapter 3), keeping these five steps in mind. This time try to follow the illogical logic of my controlled mental flight of fancy as it simulated a creative thought pattern that could help bring out the Edison and Einstein in anyone.

These exercises are limited only by the imagination. Because our headquarters faces Boston's Public Garden, it's a wonderful place to send people out for a break with a minipurpose. I once asked a brainstorming group's members to use their lunch hour to find something in the park that struck their fancy.

I assumed they'd connect to the swan boats or the *Make Way for Ducklings* statues. And, while some of them did, one member of the group returned with a panhandler who briefly joined the session to give his opinions on our topic. It was brilliant, creative, and it definitely stimulated all kinds of divergent thinking.

Music is a great creativity stimulus. Artwork with lots of colorful pictures and collages can be explosive. Nonverbal exercises tap other parts of the brain.

If you are adventuresome and are intrigued by this concept, explore some of our examples in Appendix B. Try one at your next brainstorming meeting. You might be pleasantly surprised.

[i n n o v a t i o n   f u e l]

- Don't expect to conquer your invention task focus in one meeting.

- Optimize the output of your Invention phase by scheduling sessions that allow your group to benefit from the Overnight Effect, from which people are able to make new connections because their minds enter a state of relaxed concentration overnight.

- Use a brainstorming process throughout Invention that allows plenty of time for divergent and convergent thinking.

- Start your brainstorming by presenting your task focus, followed by your answers to the five questions, to avoid the Information Overload roadblock. Trust the power of these questions and your Discovery nuggets to provide your Invention team with sufficient information to get it started on idea creation.

- Encourage people to shoot for the stars with their ideas. Create an atmosphere where people can throw out wild and crazy notions without fear of having their ideas hit by bazooka blasts. Be sure to model the behavior you want by offering up some wild ideas yourself.

- Use a variety of techniques that simulate the thinking processes of great thinkers to prompt more idea generation and more reaching for the stars. For instance, use controlled mental flights of fancy to take people's minds to another place so they can make new mental connections that will prompt more wishes.

CHAPTER THIRTEEN

Invention (Part 2)— Convergence

Once you have processed all of the Discovery nuggets in Divergence and have drained every last possible wish from team members through a variety of creativity-stimulating techniques, you will begin the Convergence portion of your Invention phase. Here the team and the champions (i.e., decision makers) will begin to select and develop from this abundance of beginning ideas to determine those you will take with you into the next phase of innovation.

This selection and development of ideas is your most important step in the entire innovation process. This is true because it is the stage at which any innovation initiative is most vulnerable because the selection and development process relies totally on the team's and its leaders' abilities to select for newness, while ignoring (or at least minimizing) the need for feasibility.

This is where the Newness/Feasibility Schizophrenia innovation roadblock arises. You will learn a lot about this roadblock in this chapter. As you read, keep in mind that this key success factor—making the leap of faith during Convergence by selecting strictly for newness and putting feasibility concerns aside for the time being—is the one that truly opens the door to innovation breakthroughs.

During Convergence, you will move through three steps:

1. Selection of wishes and themes for development

2. "We could" exploration of wishes and themes
3. Reflective, open-minded evaluation of beginning ideas

Step 1: Wish/Theme Selection

Assuming that you really do want to go for breakthrough newness, your team's ability to achieve this goal will be greatly enhanced if it has already developed skills and experienced success working with and developing potentially breakthrough or transformational nuggets in its Discovery sessions. Knowing how to evaluate wishes in a way that does not kill off newness is essential if any truly innovative ideas are to be left standing at the end of the Invention phase. This capability comes into play both in the selection of wishes and as you then focus on how to move these selected wishes toward feasibility. But, before developing potentially breakthrough wishes, you must first identify and select them and that's the big challenge.

Why Wish Selection Is So Critical

One of the reasons that the richness behind so many potentially breakthrough wishes disappears at this point of selection is that the team doesn't understand that there is more work to be done with the ideas and themes that are embedded in most wishes. This work must be completed first before determining if the ideas embedded in those wishes are worth pursuing or not.

Too often, companies *say* that they want breakthrough, new ideas from their innovation teams. But what they too often *mean* is that they want that newness to be both breakthrough and implementable without risk. Unfortunately, it doesn't work that way, particularly at this first selection point.

The Newness/Feasibility Schizophrenia Roadblock

This is where the facilitation, leadership, and idea evaluation skills become critical. If the team's champions say they want true newness and then show by their actions in choosing ideas that they are only willing to opt for ideas that are not truly new, the innovation effort will fall flat on its face. Team members will feel they have been misled about the purpose of their work and will retreat back into the safekeeping behaviors that discourage risk taking and creativity.

As I have argued in previous chapters, breakthrough innovation is largely unmeasurable. This is particularly true in these early stages of Invention, especially at this point when the group is asked to select fragile, beginning wishes for further development.

I used to be surprised when this moment arrived in brainstorming and the team leader would ask team members to nominate ideas for further development based on criteria like:

- Potential market size
- Potential return on investment
- Potential share of market
- Potential profitability

This poses an almost impossible challenge to any group of ideators. How could people possibly have a clue about these highly quantitative measures when evaluating a wish, especially after they've been encouraged to stretch beyond their normal comfort zones throughout the Divergence portion of the work? If they've done their work well in Divergence, they will have produced any number of wishes that are far beyond the realm of incremental improvements on existing products or services. Again, intro-

[FIGURE 13.1] Convergent Thinking

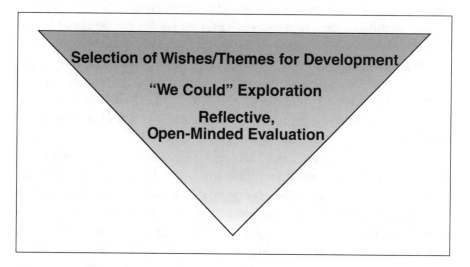

ducing a demand for quantification of possible financial outcomes at this point is like jamming on the brakes right after accelerating to 120 miles per hour, and it creates an almost schizophrenic mental reaction.

Selecting Wishes for Development

Here's another real-life scenario: You have spent an entire day digesting all of the output from your Discovery sessions and generating hundreds of wishes that directly and indirectly address your task focus/BHAG. These wishes are captured on flipchart sheets or in typed technographer notes, and they range from dull and predictable to absurdly speculative and hugely exciting. Before the group leaves for the night, you ask them to participate in a wish-selection exercise.

Everyone understands that there are more ideas embedded in those wishes than any group could possibly mine, so the challenge is to help the team champions (leaders) in choosing the highest-potential wishes that will be addressed tomorrow (or in the next session).

I like to use an hors d'oeuvre metaphor I picked up years ago when instructing a group at this critical step. I hold up a pad of paper, simulating a tray, and say something like, "Please imagine that this is an empty tray and every wish you've generated is a potential hors d'oeuvre that might fit on this tray. Your job, in this selection exercise, is to identify every candidate hors d'oeuvre that you believe should be on this tray."

When asking a group to select the most exciting wishes to put on the tray for further development, I instruct them in two steps. First, we look for the exciting incremental ideas. Then, we look for the potential breakthrough ideas.

Using either different-colored sticky dots on flipcharts or colored pens on typed sheets, they are asked to first put one color of check mark or dot next to every exciting, incremental idea they can find. We sometimes call these the just-do-its because they have elements of both newness *and* feasibility.

Then, after all of these potentially incremental ideas are identified, we instruct them to now mine the remaining wishes for the potential breakthroughs. Breakthrough wishes should be selected for further development based on one criterion only: the potential for breakthrough newness. This tends to fly totally in the face of the behavior that most people exhibit, which is to select ideas according to their feasibility. But it is an unfortunate fact of life that newness and feasibility tend to polarize when it comes to potentially breakthrough new ideas. In other words, the newer and more breakthrough the idea is, the harder it will be to make it real.

This was vividly reinforced by my colleague, Frank Hines, when he analyzed an array of ideas generated in a client's new product ideation session. The results are shown in Figure 13.2.

The learning here is quite simple. In this group, the Invention team was asked to evaluate ideas generated in Invention based on two criteria: (1) newness, and (2) feasibility. The broken line tracks the team's responses on newness for each concept and the solid line tracks its responses for feasibility. As you will note:

- The ideas that initially ranked highest in newness also ranked lowest in feasibility, while
- the ideas that initially ranked lowest in newness ranked highest in feasibility.

[FIGURE 13.2] High Newness Usually Equals Low Feasibility

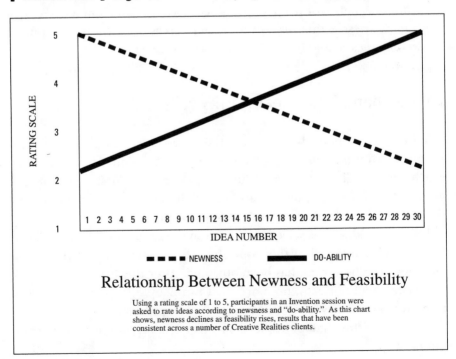

Relationship Between Newness and Feasibility

Using a rating scale of 1 to 5, participants in an Invention session were asked to rate ideas according to newsness and "do-ability." As this chart shows, newness declines as feasibility rises, results that have been consistent across a number of Creative Realities clients.

Source: CRI client study

Each time we conduct this analysis, the results are similar and validate the Newness/Feasibility Schizophrenia roadblock. The natural tendency of most people is to evoke their safekeeping side and gravitate toward ideas that are high on the feasibility scale. The problem with that approach, when the end goal is breakthrough innovation, is that *newness and appeal cannot be injected into an easily doable idea.*

In contrast, a new and appealing wish, which may at first appear totally unfeasible, can eventually become possible to achieve after some developmental work. It's just not our normal tendency to think that way and (back to Machiavelli) until individuals and teams have had experience with this, they tend not to believe it.

Because it is impossible to inject newness into an already feasible idea, the only choice is to try to work feasibility into a new, seemingly undoable idea. The challenge, and it's a huge one, is to motivate the team at this critical moment of selection to choose wishes for further development based solely on the criterion of newness or potential for breakthrough. The job of the facilitator and of the team champions is to encourage people to let go of their natural desire for feasibility and go for wisps of ideas that rate high in newness and appeal. Team champions can best do this by modeling this behavior in their own selections. If team members see that the leaders are selecting wishes with high degrees of newness and appeal, but low degrees of feasibility, they will follow this example.

Where Championed Teamwork Really Kicks In

This selection point is also where the decision-making pendulum can kill any innovation initiative because it is where decisions are made about which wishes will be developed further by the team after everyone has given their selection input. It is also where the role of the facilitators and the Championed Teamwork model become critical.

I like to conduct this selection process in the evening of the first day of brainstorming after all of the wishes have been wrung out of the group and their selection input has been obtained. After the group leaves, I ask the team leaders (ideally, no more than two people) to stay late to work with my team on the final step of this selection process.

The goal of this champions/leaders-only selection session is to generate three lists:

1. *Incremental beginning ideas.* These are semidefinable concepts from one wish or a group of wishes that share a theme with elements of doable newness that we haven't thought of before or have thought of but have never pursued, usually because of one or more seemingly insurmountable hurdles.

2. *Potentially breakthrough or transformational wishes.* These are unusual wisps, themes, and beginning directions for innovations that could revolutionize the business or marketplace.

3. *Parking lot ideas and wishes.* These are off-strategy ideas that should not be lost.

Vesting the decision making for this selection in a team's leaders, after they receive input from the group, is the first step in achieving that paradox of Championed Teamwork, which is so important for truly new ideas to move forward. Good leaders will always respond to the energy of their group by selecting a number of ideas with multiple dots or check marks from the group. This makes sense and we always encourage it.

And, while we believe passionately in the power of teamwork for the successful implementation of any form of business innovation, we have also come to recognize and respect the power of the visionary idea champion. This means that we will spend a lot of time during this selection session encouraging the team's leaders to select some ideas that tickle their fancy but might only have one or two dots next to them (probably their own). In this way, the group gets to work on its favorite ideas and the leaders are able to harness the power of the team's thinking skills to explore their educated gut. This models the balance between autocracy and consensus that the facilitator is able to provide in the Championed Teamwork model.

There is another important step in achieving Championed Teamwork and that is the provision for an idea champions hour. At selection, certain individuals in the group may see something in a particular wish that the entire group misses. Many of the ideas with only one dot might be some of those. To provide an effective balance between the group and the individual, allow for a period of time at the end of Invention where individual idea champions can stand up for their favorite ideas. Let them contract with the group and its leaders for ways to keep those ideas alive, along with the group-generated and leader-sponsored ideas.

Employing these multiple forums—group input, team leader selection, and individual idea champions—helps achieve the goals of Championed

Teamwork and avoids the negative impact of the decision-making pendulum, as explained in Chapter 5.

Step 2: "We Could" Exploration

As mentioned above, one of the ways to motivate the group to select wishes based only on newness and breakthrough potential is to promise that, before any go/no-go decisions are made, this selection process will be followed by developmental thinking exercises. This helps encourage the group to suspend its need for feasibility and go for newness because it knows that ideas will be further explored for answers to feasibility concerns.

But the people who take the leap of faith must be rewarded with follow-through on the promise of further idea development. Figure 13.2 describes how to do that.

[FIGURE 13.2] **"We Could" Exploration of Breakthrough Wishes**

---

1. Hold onto the list of just-do-it ideas until this exercise is finished.
2. Take any one of the potentially breakthrough wishes or themes and ask the group to generate as many ways as they can dream up for making that wish or theme real. Encourage a simple linguistic device that begins these ideas with the phrase "We could . . ."
3. Consider conducting one of those controlled mental flights of fancy during this time to stretch the group to come up with their most creative "we could" ideas.
4. After you have wrung every possible "We could . . ." out of the group, call a break and ask either the team leader(s) or the idea champion(s) to summarize everything he or she has heard into a beginning idea, in this format:

Beginning Idea

Code name:
Concept summary: "I'm thinking of doing <u>*(headline of beginning idea or*</u>
<u>*strategy)*</u>_____."
List of key idea/strategy elements:

_____

_____

_____

Include as many defining elements as possible. This is critical to making sure everyone truly understands the idea. For example, imagine that you are part of a team that is inventing new human resource concepts for retaining skilled staff. One concept you might develop could look like the idea in Figure 13.3.

Now, your beginning idea is ready to be evaluated. Repeat this process for all of the ideas that you've selected for further development.

Step 3: Reflective, Open-Minded Evaluation of Beginning Ideas

Before we examine the right way to evaluate and develop ideas, let me describe the wrong way, which, unfortunately, is the method that dominates in most organizations. This is the final behavioral speed bump mentioned in Chapter 8—reactive thinking. If you don't remove this speed bump, all your hard work in Discovery and idea generation in Invention may be for naught.

Reactive thinking happens because, when evaluating new ideas, most people tend to look for flaws first and operate with an instant on/off switch. This is particularly true in business environments. Using a reactive form of

[**FIGURE 13.3**] **Beginning Idea**

Code name: Operation "Summers Off"

Concept summary: "I'm thinking of shutting down the company for the entire summer following every year that we achieve our goals."

List of key idea/strategy elements:

- Contingent upon meeting preagreed measurables

- At discretion of partners

- Use part-timers to maintain client service

- No carryover or accrual ("use it or lose it")

- Maybe just for certain levels of the organization

evaluation, we seek to make immediate and quick decisions. Ideas are either good or bad, valuable or worthless, go or no-go. This type of evaluation deals only with the facts as stated, not the intent or the possibilities behind an idea.

In such an atmosphere, unless an idea is completely perfect, people will immediately start looking for reasons to hit the off switch: "The cost of this idea will be too high." "We couldn't patent this." "Management wants to go to market in six months and this idea is too complex for that." And on and on until the idea suffocates and dies under the weight of the negativity heaped upon it.

Contrast this way of thinking with something called reflective or open-minded evaluation, in which ideas are considered with an open mind. This approach to idea evaluation helps individuals and whole groups to understand the intent as well as the statement, the possibilities and not just the facts. In a culture of reflective, open-minded evaluation, the response to a new, but flawed, idea is "Gee, that's an interesting beginning idea; let's focus on how we could fix what doesn't work about it."

The reflective, open-minded idea evaluation process follows these six steps:

1. Review your first beginning idea, as defined in the "we could" exercise, with the team.
2. Start evaluating the idea by inviting team members to list everything the idea has going for it (the pluses). This will take some getting used to because it's the reverse of our normal way of evaluating ideas, which is to first point out their flaws. By starting with a list of the idea's strengths, you will accomplish two important things. First, it will help everyone see what others see that supports the idea without selling. This is very important, because invariably, everyone learns something that the idea has going for it, whether they liked it already or not. One of those pluses may turn someone's initially negative reaction to an idea into a positive attitude toward it. The second thing the pluses do is to give the idea momentum and weight to survive the negatives.
3. Express issues with the idea by using another linguistic device, the phrase "How to. . . ." List major stoppers—concerns that would stop the idea in its tracks. Using this clever invitation to problem solving automatically invites solutions. Examples of the type of thing that might come up are "How to make this affordable" or "How to do it in less time."

4. Problem solve each how-to issue by inviting suggestions for creatively resolving it until you have successfully overcome it. Go on to the rest of the issues, one at a time, until you have addressed all of the issues that need to be overcome. Be prepared to modify your idea, where necessary, to accommodate creative solutions to the issues.

5. Restate your idea after the problem solving is complete, including all modifications resulting from the open-minded exercise.

6. Test the idea against the three threshold criteria for moving forward: (1) Is it new? (2) Is it feasible? (3) Am I committed to a follow-up action?

The idea that gets a yes to all three of these final questions should then be moved forward to Phase 4, Greenhouse, where it will undergo further development. The idea that gets a no to any of these three final questions is not ready to move forward and either needs more work, should be put on hold, or should be discarded.

It is very important that the team and its leaders be prepared for the wish that never makes it to Phase 4, because this should be celebrated, not viewed as a failure. It means that the team had the courage to go down a dark tunnel in its relentless pursuit of newness, but didn't find the bright light. To paraphrase Thomas Edison, from my earlier light bulb invention story, "Now you know a solution that won't work." That's cause for celebration, not for self-flagellation.

The great thing about engaging in developmental thinking in Convergence is the more time you have, the more wishes you can explore. That is why we insist on at least an additional day for this Convergence portion of Invention.

Now that you have successfully developed one wish and the group understands the process, the team can be broken out into smaller subsets to tackle this process of generating "we could" thoughts, beginning ideas, and reflective, open-minded evaluations on as many wishes as possible, including that just-do-it list that you've had on hold. Normally, those ideas can skip the "we could" exercise and go straight to reflective, open-minded evaluation.

Big Success with a Less-Than-Perfect Idea

To support and hopefully validate this discussion of wish selection, development, and evaluation, here's a classic, well-known story that demonstrates what can happen when you are willing to devote time to a less-than-perfect idea. In the early 1950s, Lionel Alexander Bethune Pilkington, a British industrialist whose family was in the glass business, was looking for a way to eliminate the need to grind and polish plate glass.

One night, while washing dishes, he began to notice an oil slick forming on the water. It was very smooth and the oil sat perfectly on top of the water. Imagine his surprise when this wish popped into his head: "I wish I could float molten glass on water."

Imagine the absurdity of that idea. Consider what would happen to you if you offered up a silly idea like that in your world. You'd be laughed out of the room.

Unlike most people, however, Pilkington didn't just dismiss this idea; he didn't censor himself simply because the idea was seriously flawed (because molten glass could never float on water).

Following the open-minded evaluation model I described above, he instinctively took himself through those steps that we like to teach. He began by extracting the key principle of the idea that appealed to him, which was the concept of floating.

Next, he did something critically important by using that simple how-to language I mentioned above. Most people would have said, "Pilkington, you silly twit." (He was British, after all.) "Everyone knows you could never float molten glass on water. Get serious!" Instead, he asked himself something like, "How to float molten glass on something that could cool to room temperature and result in a perfectly smooth piece of glass that wouldn't have to be polished by hand labor."

He spent the next seven years experimenting with hundreds of materials on which he could float molten glass. In 1959, he obtained a patent for the float glass process by which molten glass is floated on molten tin. This is now the method by which most glass is made today. And, like so many innovative breakthroughs, it seems simple in hindsight.

In addition to illustrating the power of flawed ideas, Pilkington's example also points out another important thing that innovators must understand. Most of the ideas you're going to come up with are actually composites of a number of little ideas.

For Pilkington, his idea broke down into two components: the idea of floating molten glass on something and the idea of what that medium would be. Initially, he had the medium wrong (water) but eventually he got it right. Like Pilkington, you can learn to break your ideas down into their various parts and see which parts work and which don't; then work on solving the problems of the parts that don't work instead of rejecting the whole idea as being unworkable.

A Winning Idea That Evolved from Failure

Here's another example of the power of this staged approach to what I like to call the seasons of idea generation and development. It also points out how ideas that initially appear to have failed can be combined with other beginning ideas to create a winner.

This story involves a project I facilitated for inventing new restaurant concepts for that theme park client I've mentioned in a couple of earlier stories. Several years ago, the client opened a new part of a theme park and quickly learned that it was so popular that it needed to be immediately enlarged. Concepts for new attractions, restaurants, and concessions all needed to be created and built as quickly as possible.

Thanks to a man named Kevin Malchoff from Rich Products (another client) I was introduced to the man who was then head of all food service within the park. This person invited Creative Realities to help him create new restaurant concepts for his theme park.

With my coaching, this client invited about 15 diverse park employees to participate in Invention meetings at a remote location on the park property. During their Divergence wishing session I ran two controlled flights of fancy. The first one asked the group to remember favorite old TV shows. One of them was *Happy Days* and the image of Arnold's Diner emerged, where the waiters and waitresses skated to parked cars on roller skates. This led to a beginning restaurant concept where guests would travel in cars through a diner.

Unfortunately, (or fortunately, really) this idea stalled during its reflective, open-minded evaluation, because the idea was becoming more of an attraction than a restaurant that could be run cost-effectively.

The second controlled flight of fancy involved an exploration of vanishing Americana. One of the items listed in this exercise was drive-in movies and it led to another round of developmental thinking around the idea of drive-in movies. This one also stalled because the drive-in's snack shop just wouldn't be able to generate enough sales volume.

Now, here's where the breakthrough came. Because the team understood that pieces of ideas can be combined to form new ones, they lifted the idea of Arnold's skating waiters and waitresses and planted it into the drive-in movie setting. The result is a restaurant, where wait staff brings food to guests who are "parked" in cars that are tables, while fun movie trailers play on an "outdoor" movie screen.

Here's a perfect example of how teamwork and perseverance can help any group to achieve real newness even though, at first, it seemed as though the ideas were failing. This is what it takes to nurture an innovative culture.

[i n n o v a t i o n   f u e l]

- Do not shortchange your innovation effort by imposing quantitative criteria for idea selection. This is not the time to ask how big the market might be, how profitable the new product might be, or similar quantitative questions.

- Develop the skills needed to take the innovation team successfully through the three-step process of Convergence—selection of wishes and themes for development, "we could" exploration of these concepts, and open-minded evaluation of your beginning ideas.

- Make sure everyone understands that the closer you come to true newness, the further away you generally move from feasibility. Help people accept that this scary reality of innovation can be overcome by mastering the right skills for open-minded idea evaluation and development.

- Have the courage to eliminate the Newness/Feasibility Schizophrenia roadblock by choosing ideas with a high degree of newness instead of those that are immediately feasible. Make sure the team leaders model this behavior for the rest of the group.

- Get input from all team members about the ideas that they favor, but then use the power of Championed Teamwork to propel truly new ideas forward. In a champions/leaders-only selection session, generate lists of incremental ideas, potentially breakthrough ideas, and off-strategy ideas that have merit and should not be lost.

- Promise team members that selected ideas will be further developed later in Convergence and then fulfill that promise with open-minded idea development and evaluation.

- Explore breakthrough ideas through open-minded evaluation, which starts by listing the pluses of an idea, then goes on to idea stoppers, and answers them with problem solving using a how-to construction.

[PHASE FOUR]

Greenhouse

CHAPTER FOURTEEN

Making Your Seedlings Bloom

Now you're ready for Phase 4 of the innovation process—Greenhouse. You enter this phase with great ideas, which are still only concepts. In Greenhouse, you will undertake all the internal development, testing, and prelaunch activities needed to turn those ideas into realities; to get them ready to be released into the marketplace. If your innovation is a new product or service idea, your work in the Greenhouse also may include some test marketing.[1]

As you move from the Invention phase into Greenhouse, no doubt your innovation team will be pumped up about the breakthrough potential of the innovative ideas you have developed. This excitement, however, may be tempered by worries about what lies ahead, especially if you've gone through other innovation efforts where good ideas died aborning. This is the beginning of that Dark Night of the Innovator I've mentioned so many times. It's called that because it is here that great ideas too often go to die. It's where the difference between creativity and innovation becomes clear.

But, take heart. Unlike previous efforts, where you probably got bogged down and ran into numerous speed bumps and insurmountable roadblocks as you tried to negotiate your ideas through to reality, the stage is now set for success.

To ease your fears about failure, remember that you are approaching innovation in a new way this time. By following the model I've provided, you have spent sufficient time in Phase 1, Setting Objectives, to make sure you have a core purpose, core values, and BHAG/task focus that everyone understands and supports. You have also done everything possible in Phase 2, Discovery, and Phase 3, Invention, to develop innovative ideas that fulfill that BHAG/task focus. By doing these things, you have negotiated your innovation effort around numerous roadblocks, including the truly devastating ones, like the Garbage In/Garbage Out pitfall and the Leadership-Empowerment Fable.

Having devoted the necessary (some would say extended) time to the thorough and rigorous execution of the first three phases of the Creative Realities model of innovation, you are now prepared to move more quickly through the final two phases, Greenhouse and Implementation and Launch. In fact, one of the many positive by-products of our approach has been markedly faster development, implementation, and launch time lines for new, truly innovative ideas.

This is not to say that everything will be smooth sailing and you won't have to live through Dark Night of the Innovator episodes. However, the number of such trials should be lowered and your ability to overcome these obstacles increased by having established and communicated an objective that has organization-wide support and by having strong, innovative ideas. Both of these improvements are the result of the hard work you've done in the first three innovation phases.

Why Stage Gate Models Don't Work

Now let's look at what you need to do to make realities of the exciting new ideas you have developed. First, be aware that I'm not going to spend a lot of time suggesting specific processes for greenhousing, implementing, and launching new breakthrough ideas. This is the time in your innovation journey where the individual nature of every organization requires customized solutions. During Greenhouse, in particular, you will need to develop your own road map, for every idea you want to pursue, because trying to follow a single, rigid methodology prescribed in a book will rarely produce the desired results at this critical stage of innovation.

Specifically, I want to issue a warning about using a Stage Gate model, in which gangs of people assemble at preset time intervals to make decisions on an array of innovative ideas that are all vying for developmental dollars and human resources. In any of the models I've seen, decisions tend to be made against a one-size-fits-all set of evaluative criteria by a committee. Under this model, as any truly innovative idea evolves, it must successfully pass through a series of decision gates, the end point of which is, theoretically, a successful marketplace launch or implementation.

While this methodology can be used throughout the whole innovation process, its primary purpose usually kicks in after idea creation, when organizations try to drive fragile new concepts through to reality. I believe Stage Gate processes have evolved as a well-intentioned, but wrong-headed response to the desire to systematize innovation in order to increase success rates. But, as I've said before, innovation is inherently messy and resistant to systemization.

My experience in watching companies trying to follow Stage Gate models tells me they simply don't work very well. Here's why:

- Stage Gate models, almost by definition, require ideas to be pushed uphill through layers of doubt. With this model, virtually any functional stakeholder who just doesn't "get" an idea can slow it down or even kill it directly or indirectly. In worse-case scenarios, Stage Gate models allow individuals, who believe a new idea goes against their self-interests or perhaps the interests of their departments, to use the nature of the model itself to block success.

- Because any new idea must pass through many layers of functional stakeholder decision making, Stage Gate models foster a lowest-common-denominator form of consensus. This invariably strips breakthrough newness out of any idea because, in effect, it fractionates the decision-making power between a series of individuals who make their decisions under less-than-optimal circumstances.

- Stage Gates attempt to institutionalize layers of decision making on a bottom-up process. As explained in Chapter 6, it is only possible to empower a bottom-up innovation model if the idea-generating team is just one level removed from the decision maker(s). Following this alternative, in effect, eliminates the need for Stage Gates.

- Stage Gate models assume a fairly predictable developmental flow for virtually every idea created in an innovation initiative. As explained throughout this book, ideas don't follow one clear path to success. Therefore, no one developmental model for decision making can exist when you are pursuing innovation.

- The mechanism for decision making in any Stage Gate model is rooted in being able to quantify the likelihood for success of the idea you're working with because this is the only way to convince so many decision-making participants and influencers. Again, as I've discussed, breakthrough innovation cannot be quantified in its earliest developmental stages.

- Stage Gates tend to reinforce the Garbage In/Garbage Out roadblock by sustaining the myths about the unpredictability of the fuzzy front end of innovation. So far, all of the Stage Gate models I've seen ascribe an almost black-box attitude toward the process of idea creation. They acknowledge a place for it but don't provide sophisticated mechanisms for how to do it. They then institutionalize an almost freight-train mentality around idea development, reinforcing quantity and procedure over quality.

Unfortunately, in their attempt to manage creativity and innovation, Stage Gate models substitute form for substance. They provide an A-plus-B-plus-C procedural flow that allows for too much innovation cop-out. Instead of working hard to overcome roadblocks and speed bumps, people can stick with their old ways and then, when the innovation effort collapses, they can claim, "But we followed the process!"

The reality is that after they are tried and seen to fail, because of the weaknesses outlined above, Stage Gate processes for the development and implementation of innovative ideas end up as book ends and office doorstops.

Key Success Factors for Greenhouse

The critical element missing from Stage Gate models is a full consideration and understanding of the people part of innovation. Given the messy and unpredictable nature of innovation, the challenge is not to create a clean, predictable process. The challenge is to introduce skills and procedures that will help people deal with the ambiguities and paradoxes that are part of every innovation effort.

So if Stage Gates and similar rigid processes won't help in Greenhouse, what will? Here are the keys to success in Greenhouse:

- An array of potentially breakthrough ideas
- A Championed Teamwork decision-making protocol that fosters organization-wide teamwork and provides for consensus of the I-can-live-with-it kind and avoids lowest-common-denominator decision making
- A visible strategic vision (core ideology and envisioned future) including an innovation task focus as a touchstone for guiding the whole organization in support of the innovation initiative
- An open-minded, developmental mind-set versus a reactive (yes/no) mind-set
- Effective ways to expose fragile beginning ideas to the customer-consumer spectrum with the goal of idea development versus go/no-go decision making
- Iterative processes that foster how-can-we-do-this? thinking versus an evaluative attitude during technological development
- A clear and consistent marketing/communications/branding strategy (for product and service innovations)
- A rigorous assessment of potential competitive response(s) to your innovations

You'll notice that some of these success factors are the same as ones you read about in earlier phases, but they are important in the Greenhouse phase for slightly different reasons as you'll learn in the next few pages.

When to Get Senior Management Buy-In

As mentioned already, I no longer advocate that senior management decision makers be involved in every step of an innovation program. However, there are three times when the involvement and buy-in of decision makers is critical for both idea development and organizational momentum. Note that two of these three points are closely entwined with the Greenhouse phase:

1. In Phase 1, Setting Objectives, where the strategic vision and task focus are created. This is where the all-important bond is formed

between the aspirations of the decision makers and the inspiration and empowerment of the team.

2. At the end of Phase 3, Invention, when an array of exciting, fragile beginning ideas should be reviewed and resources obtained for Phase 4, Greenhouse development.

3. At the end of the Greenhouse phase, where implementation and launch decisions are made and capital resources allocated.

Assuming that the team has engaged either in-house or external facilitators, these important interactions with senior management should be orchestrated to occur at these three optimal times for sharing information and developing hypotheses in a way that informs rather than seeks permission. These shirt-sleeves meetings are the linkages between the empowerers (decision makers) and the empowered (innovation team). If scheduled and executed properly, they can help ensure the avoidance of many innovation roadblocks.

Unfortunately, in far too many organizations, senior managers remain mystified as to why their stamp of approval spells the difference between fully supported implementation and fierce resistance to new ideas during Greenhouse. Yet, if senior management support is critical for running the established business, just imagine how much more important it becomes in supporting the infinitely harder job of innovation.

People, systems, and whole organizations resist change. It's as simple as that. Without clear and obvious support from the top, innovative ideas—particularly if they're of a breakthrough nature—will have to fight that much harder to survive. And, while a survival-of-the-fittest mentality is important for innovative ideas, most organizations underestimate the degree of resistance that great, new ideas will face during Greenhouse. It is part of senior management's role to minimize this resistance with a clear show of support for innovation.

This approach will also create real visibility for another critical signal that supports innovation: failure. Again, as hinted earlier, I truly believe that most senior management teams try to empower innovation using Mahr's Law of Limited Involvement (Don't get any on you!). Most innovation efforts fail. If organizations link innovation-related decision making in the ways that I am proposing, the result will have to be more learning, less blamestorming. Every level of the organization will appreciate the risks and imponderables that accompany every innovation effort. Expectations will be readjusted from the win/lose absolutes that have historically resulted in unfair armchair quarterbacking. And, hopefully, internal roles and responsibilities focused on the development and introduction of innovations will become career stepping stones versus career tombstones.

Championed Teamwork

Obtaining senior management's visible acceptance of ideas at key junctures, especially immediately prior to Greenhouse and at the end of Greenhouse, is closely tied to the Championed Teamwork decision-making model introduced in Chapter 5. As you will recall, the goal of this model is to provide the facilitated balance between autocracy and consensus that will allow innovation team members to provide idea-shaping and development input without watering down the newness of ideas.

Clearly, in the Greenhouse phase, championed teamwork becomes critical to the survival of new ideas as fledgling ideas are exposed to a broader spectrum of critics within the organization. Those evaluators who did not participate in the first three phases of innovation will tend to question everything your team has done. Having a well-established teamwork model that allows skeptics to be heard, but at the same time doesn't let them stall forward momentum of ideas, is essential.

Strategic Touchstones for Decision Making

One of the ways to help engage those who haven't participated in your earlier innovation work and lessen their skepticism is by communicating the strategic touchstones (core purpose, core values, BHAG/task focus, and vivid description of the future) that guided innovation-related decision making right from the start. Failing to communicate this key information is a major reason many innovation initiatives run into roadblocks during the Greenhouse phase. People throughout the organization who are not directly involved become confused or feel threatened by the hints of newness that inevitably begin to leak out from the work of the innovation team. Such fears or discomfort tend to be further exacerbated if people have no context by which to evaluate what they're hearing via the grapevine.

To prevent such problems, senior management decision makers must effectively communicate those touchstone elements in consistent and highly visible ways. Ideally, this should happen at the end of Phase 1, Setting Objectives, before the Phase 2 Discovery work begins. But this is not always necessary, especially in those industries where ideas have a short half-life and where turnover tends to be high. For these reasons, it is sometimes wiser to wait until the start of Phase 4, Greenhouse, to share these goals with the whole organization.

No matter when this communication takes place, the success of the Greenhouse phase is maximized when the whole organization understands at least the context for an innovation initiative. Effective communication of this context makes it more likely that everyone will get behind ideas that, without this context, may seem absurd, counterintuitive, or just downright stupid. Visible affirmation of the organization's strategic reason for being, its core values, and innovation goals will also help key implementation stakeholders better understand, especially at those times during Greenhouse when they may be asked to do something at which they might otherwise balk.

Keeping an Open Mind

I've already talked several times about the difference between open-minded evaluation and reactive thinking. Being able to explore ideas without rushing to premature judgment about their feasibility and without trying to quantify their chances of success too early in the developmental process remains critical in Greenhouse just as it was in Discovery and Invention.

Some of the obstacles that typically arise in Greenhouse, when an organization is not able to embrace open-minded evaluation techniques, include:

- Giving up on an idea as soon as it encounters negativity on the developmental highway
- Pride of authorship that fights against any modification to an idea as it is being developed
- A tendency to modify an idea so dramatically that it loses its uniqueness and its connection with the strategic touchstones

Be alert throughout Greenhouse for signs of these obstacles. Remain true to the skills of open-mindedness to help the newness in your ideas to remain alive. Here, again, is where the role of the facilitator becomes crucial. Good facilitators are able to keep the team focused and can help reinforce that how-to attitude, which becomes so important to the successful weathering through the Dark Night of the Innovator.

Greenhouse and Consumers: Idea Development versus Permission

Open-minded evaluation is essential to protecting ideas from internal threats during Greenhouse. But Greenhouse is also where consumers will kill your potentially breakthrough ideas if you let them. This is because Greenhouse is the place to explore the different ways you might communicate your new ideas. Unfortunately, many organizations turn it into a place where consumers are allowed to make yes/no decisions. Let me explain.

Think about a great new idea that you first resisted. I can think of two terrific concepts that, at first, held absolutely no appeal for me:

1. Debit cards
2. Hertz #1 Gold Service

Debit Cards

When I was first exposed to the concept of a card that would automatically extract cash from my checking account, I couldn't imagine why I would be interested in something that wouldn't provide me with the 30-day grace period that my credit cards provide. I thought it was a dumb idea.

Today, my debit card is my favorite card. I use it every day, and virtually never go near a bank branch or ATM machine. Now, $200 in cash lasts forever. I am becoming virtually cashless. Yet somehow, when I was first exposed to the concept of a debit card, it wasn't presented to me in a way that helped me "get" it.

Hertz #1 Gold Service

As a traveling consultant, I am on planes and in rental cars constantly. Naturally, I was a prime promotional target when Hertz launched its breakthrough new #1 Gold Service. This wonderful idea, where a personally dedicated rental car is running in a parking spot with my name on it, ready for me when I step off the Hertz bus, with my car's location displayed on a big sign, was tailor-made for me.

Unfortunately, what caught my eye in the initial promotional mailer from Hertz was the legalese about my liability for any car that someone might claim with my driver's license. I rejected the initial offer. But then I got upset with Hertz when I stepped off their bus one day and saw row after row of fresh, shiny cars, gassed and ready to go for other people. Of course, I signed up that day.

Idea Development versus Permission

Too often ideas are exposed to consumers in the Greenhouse phase in a way that seeks a yes/no response instead of a more developmental response. This is where that interactive conversation technique called Co-Labs (which I introduced as a Phase 2 Discovery tool in Chapter 10), comes into play again. Conversations with consumers can be powerfully developmental if used properly here in Greenhouse.

For example, I can imagine a different conversation about debit cards than the one that happened when I was first exposed to the concept. What if it had begun by exploring the relationship I had with my bank at the time? What if someone had asked me what I was putting up with?

I might have mentioned how frustrating it was to find an ATM that was out of service when I most needed it. I might have been able to reflect on the number of times I needed to use one every week. I might even have wished for something that was embedded in the concept to which I was about to be exposed.

Let's also assume none of that happened. Instead, what if someone had worked with me until I really understood the concept and then asked for my help in communicating it to other nonfinancial minds like mine? What might have happened? Instead of walking away not understanding the idea, I might have become the evangelist for it that I actually became once I finally understood the concept on my own.

Turn the traditional focus group, in which consumers are asked if they like or don't like an idea, into Greenhouse discussions where consumers participate in your idea development. Don't attempt to quantify anything, yet. You will come out with much stronger ideas and will have a good sense of how to communicate the benefits that will strike closest to your consumers' hearts at launch time.

Technology Myopia and Other Hurdles

In addition to having effective processes for exposing your beginning ideas to consumers, you will also need ways in Greenhouse to expose your fragile beginning ideas to the dangers that arise during technological development. Roadblocks can arise as you begin to draw on talent from your organization to conquer the technological challenges entailed in your ideas. Two of the most common ones are Technology Myopia and Patent-Protection Paranoia.

Technology Myopia results from the issue I discussed back in Chapter 5—too heavy a reliance on experts. Hopefully, you will have overcome this problem in the first three phases of innovation by forming an innovation team comprised of both technical expertise and technical naïveté. But now you have reached the stage where you most likely will have to reach beyond your core team and work with parts of your organization where the slogan is likely to be "Expertise Rules."

If the technological experts you are relying on to make your idea a reality don't get it or, in some unfortunate situations, don't even want to get it, your idea can die an ugly death.

Take, for instance, the story of how many of the engineers at Polaroid Corporation first reacted to the idea of the I-Zone Instant Pocket Camera. As reported in *The Wall Street Journal,* a three-year battle ensued between the company's scientists and engineers and the marketing people who supported the idea of a camera targeted at teenagers. The engineers argued that the Pocket Camera, which features a cheap lens that produces fuzzy, thumbnail-sized photographs, would tarnish Polaroid's reputation for creating "elegant products based on high technology."[2]

Fortunately, Polaroid's chief executive, Gary T. DiCamillo, supported the concept. Less than three months after introduction, the Pocket Camera was the best-selling camera in the United States, producing more than 25 percent of the company's new product revenue for all of 1999, even though the product was only on the market for the last quarter of that year.

As this story so vividly shows, Greenhouse is where the support of senior management and the strong communication of the strategic vision and BHAG/task focus touchstones must come into play. And, always remember that resistance to new ideas is never restricted to any single functional area of an organization. For every story where it's technology villain versus marketing hero, there's one that exhibits marketing myopia version technology hero. No matter which arm of the organization might be resisting, the message must be perfectly clear about two things if a truly new idea is to have a chance for success:

1. Management wants this idea to work.
2. This idea supports the strategic vision of the whole organization (or department or division).

Engage your functional resistors, whoever they might be, in open-minded evaluation exercises for idea development. They cannot be allowed to just throw up their hands and say, "This won't work."

Here is where the power of analogous worlds becomes crucial. By inviting experts from other noncompeting worlds who face similar problems to yours (like the dry cleaner who helped the photo processor come up with new ways of managing chemical waste disposal), you create a stimulating, non-threatening forum for creative problem solving and technology transfer.

Patent-Protection Paranoia, the other issue that often comes up during technological development, is much more of a problem in some industries (and in some companies) than in others. You may not have to deal with it in your organization unless your company has a tradition of patenting. In companies with a history of patents, the tendency is to avoid ideas that can't be owned.

Ignoring a potentially breakthrough idea because of Patent-Protection Paranoia doesn't make a lot of sense, in my opinion. This will be particularly true in this new century, as the paradoxical consumer demand for breakthrough yet standardized high-tech innovations becomes the norm. In this new environment, where consumers are growing accustomed to rapid successions of technology breakthroughs, the value of patents decreases in many fields. Companies that formerly relied for years on a handful of lucrative patents for their fiscal health may find they need a new strategy. Instead, they may find that success lies in bringing a constant stream of new ideas to market, including ones that often times will, in effect, cannibalize the ideas they brought to market just a year or so earlier.

War Games: Your Idea versus the Competition

Another key activity during Greenhouse is to consider how your competition will respond to your idea when it reaches the marketplace. The tendency among many organizations is to either completely ignore potential competitive responses or to define the competition too narrowly when evaluating it. As discussed in Chapter 11, such Competitive Myopia is dangerous, and just because you considered competitive threats during the Discovery and Invention phases doesn't mean you're done with this part of your work. Now that you have a specific idea that you're developing, it is even easier to conduct exercises to expose its competitive vulnerabilities.

Here's an example of how such an exercise can work. In the early 1990s, one of my partners, Cris Goldsmith, and I found ourselves in a room that included several communication product engineers who were ignoring a major consumer demand for privacy in cordless telephones. While they acknowledged the importance of this consumer desire (which their marketing and sales colleagues were begging for), their technological history with competition was rooted in a slower pace of change. They figured they'd launch it when they were good and ready, and they estimated that would be in about five years.

On the second day of their brainstorming session, Cris broke them into four teams, each representing a major potential competitor. He called one of those competitors "Japasonic" (a "joint venture of everything you fear about the Japanese electronics companies"), and another was Motorola, which had not yet entered the cordless phone business.

His instruction to the four teams was to take our client company out of the cordless telephone business. Each team was given two hours and was provided with a separate breakout room, a facilitator, and an attack plan presentation template.

The results were astounding. By the simple act of becoming a real, potential competitor, these same people became aggressively innovative in their plans. Removing their everyday corporate hats enabled them to break out of many constrained thinking bonds. This was particularly true for the Motorola team, which was able to be the most creative, because Motorola had no embedded cordless infrastructure at that time. This allowed that team to really take a zero-based approach to its strategy.

Not surprisingly, most of these breakout teams created ways to include that much-needed privacy feature in their cordless phones but in about two years as opposed to five. This caused our client company to overcome its Competitive Myopia and to rethink its resistance to this new technology. It also resulted in the company increasing the idea's priority significantly. The incremental benefit from this exercise was that when Motorola entered the business two years later, our client was ready.

The by-product of this experience for our company was the establishment of a modular service that we build into many of our programs in which we expose ideas to competitive threats. We call this component War Games. There are various versions of this concept available from many consulting firms, but the goal of all of them is pretty much the same: to ratchet up any organization's awareness about the power of its competition, both as perceived and unrecognized. Such exercises can be used in Discovery, but they are particularly useful in Greenhouse, where any effort that helps the team become more aggressive in anticipating competitive activity almost certainly helps make an idea stronger.

Finishing School

If the idea you're developing in Greenhouse is a new product or service, one of the final steps you'll take during this phase of innovation is the development of marketing, communication, and branding strategies. One of the big stumbling blocks companies tend to run into here is failing to align all of these pieces under one integrated, well-coordinated strategy.

This first became clear to me when we started to get heavily involved in naming as an ancillary service to our bigger innovation projects. Initially, we would ask our (seemingly) sophisticated, marketing-oriented clients for the positioning they had developed for the product or service they wanted us to help them name. Their response, too often, was something along the lines of, "We thought the positioning would follow the name." It doesn't work that way because naming is almost completely subjective. Therefore, the positioning becomes the touchstone for naming, like the core purpose is the touchstone for innovation.

Frequently, compounding the problems produced by this wrong-headed approach were the difficulties that occurred when an exciting name could never be made to work properly with packaging or collateral graphics.

For these reasons, we now insist that our clients link all of the aspects of effective branding in their sessions (including product packaging, branding graphics and imagery, collateral, signage, positioning, naming, etc.). We call it Finishing School. However you choose to do it, be sure that you provide for the synergistic development of all of these communication elements in a unified manner.

[i n n o v a t i o n   f u e l]

- As you enter the Greenhouse phase, make use of all you learned in the first three phases to help your innovation team maintain its enthusiasm and overcome the challenges of the Dark Night of the Innovator when fragile beginning ideas are in their most vulnerable state.

- Resist efforts to introduce a Stage Gate model or other rigid process into this phase of idea development. Arm yourself with knowledge of the flaws of this model so you can successfully argue against relying on it. This is especially true if your goal is breakthrough innovation.

- Make sure you enter Greenhouse with senior management buy-in to many (not necessarily all) of the ideas you hope to grow in this critical phase.

- Communicate your strategic touchstones to the whole organization and continue to use the championed teamwork model to help you overcome the problems that can arise when your new ideas are introduced to a wider audience.

- Continue to embrace the skills and techniques for fostering open-minded evaluation instead of reactive thinking as your idea begins to take on weight and become more real. Make sure that open-minded evaluation is also used in overcoming technical difficulties.

- Do not let Technology Myopia or Patent-Protection Paranoia derail your ideas.

- Participate in conversations with consumers that engage them in idea development rather than asking them to label new concepts as good or bad.

- Improve your innovative idea by engaging in War Games or similar competitive exercises that expose it to the threats it is likely to meet once it's in the marketplace.

- Develop an integrated approach to all aspects of marketing, branding, and communication for any innovations that fall into this arena.

Notes

1. Bear in mind what I said earlier, that innovations don't have to be new products or new services. So when I talk about launching your new ideas at the end of the Greenhouse phase, recognize that some of your innovations may be launched internally. These internal changes may cut costs or boost productivity without your customers ever knowing you've made any changes. With such ideas, by the end of the Greenhouse phase you will have done all the work needed to put the new practices into effect in your workplace.

2. Alec Klein, "On a Roll: The Techies Grumbled, But Polaroid Pocket Turned into a Huge Hit," *The Wall Street Journal*, May 2, 2000, pg. A1.

[PHASE FIVE]

Implementation and Launch

CHAPTER FIFTEEN

Where the Rubber Meets the Road

When you arrive at the fifth phase of your innovation process, Implementation and Launch, you know you have a workable idea—an actual innovation, not just a theoretical one. Technical hurdles have been overcome. It's now time for the organization to put its money where its mouth is to implement and/or launch the innovation. By the end of Phase 5, your innovation will be meeting its public—whether that is consumers, in the case of new products or new services, or internal audiences (and maybe vendors) for new business processes or other innovations that impact only your internal operations.

Depending on the degree of newness of the innovation you are pursuing, Phase 5 can be the longest single phase in terms of time. Generally, although not always, the more breakthrough the idea, the longer its implementation takes. And, because scale is involved, this is also where you are most likely going to spend the most capital.

For instance, let's suppose your innovation is a completely new food product that requires setting up a new production line featuring manufacturing equipment that you've designed in the Greenhouse phase. Clearly, this implementation is going to take a lot more time and money than would an incremental innovation, such as adding a new scent to a shampoo, which can be out the door and on store shelves in a matter of months.

Similarly, the time required for the launch portion of Phase 5 can also vary considerably depending on how breakthrough the innovation is. For instance, as Geoffrey Moore made so abundantly clear in his best-seller, *Crossing the Chasm,* a carefully considered launch that takes into account the various stages of consumer acceptance is required to earn success for truly new high-tech innovations.

With nontechnical products, the launch period, of course, can be considerably shorter. For example, consumers don't need months of advertising to be convinced to risk $3 on a new shampoo the way they need to be wooed into plunking down $2,500 on some great new home entertainment gizmo.

What to Expect in Phase 5

The length of time you spend in Phase 5 and the complexity of your route during this portion of your journey on the innovation highway will impact how smooth this section of your trip will be. Clearly, the longer and more complicated this phase, the greater potential there is for problems to arise, for the anti-innovators to plant doubts and cause dissension, and for senior management's enthusiasm to ebb. Cost overruns, unexpected delays, market shifts, or unanticipated moves by competitors can dissipate corporate resolve and produce numerous headaches and Dark Night of the Innovator nightmares for your implementation team. (Note that just as your original innovation team was expanded during the Greenhouse phase, it will expand even further during Implementation and Launch.)

Key Success Factors for Implementation and Launch

No matter how short or lengthy your Phase 5 may be, here are the keys to success for avoiding the problems that typically arise along this final stretch of the innovation highway:

- A clear, visible, flexible, and agreed-on beginning cross-functional timetable for Implementation and Launch
- Periodic check-ins by the original innovation team with the broader implementation team, including senior management and the leaders of key stakeholder functions

- Periodic linkage back to the organization's strategic vision and innovation task focus
- An effective decision-making procedure for killing ideas, particularly those that have taken on a life of their own. A good innovation process helps teams recognize when an idea should die.
- A rigorous, disciplined postmortem assessment of every innovation effort to extract key learnings and "therefores" for subsequent innovation initiatives

Implementation Road Mapping

Many sophisticated software models exist for setting up your Implementation and Launch timetable and plan—what I think of as your Implementation road map. Be sure to choose a model that identifies the critical, interactive path between the various functions that are crucial to successful Implementation. Your objective is to develop and gain organization-wide commitment to a beginning road map that is clear, visible, and most of all, flexible.

Also be aware that too often this road-mapping exercise is conducted without consideration of the people part of Implementation. By that I mean, don't overlook the natural interaction of functions and the tendency, in almost every organization, for some of them to have a historically adversarial relationship. If you develop a road map that naïvely ignores this baggage from the past, your Implementation effort can be seriously hampered or even defeated.

Again, while you definitely need a well-developed timetable, Phase 5 is not the place for rigid protocols such as Stage Gate processes. It is where flexibility continues to rule. It is also where the resources of the organization become most scarce, depending on the number of innovative ideas being pursued. This is further exacerbated by the tendency, in even the largest organizations, to call upon the same people in each functional department to be part of every new Implementation effort. It's almost as if everyone knows, instinctively, who those few get-it-done people are. But even these masters of getting it done can find their plates too overflowing to be totally effective at times.

Three other roadblocks that typically arise when developing the Phase 5 road map are:

1. The Focus/Resources Bottleneck
2. The Bataan Death March
3. The *Marty* Dilemma

The Focus/Resources Bottleneck

Organizations develop the Focus/Resources Bottleneck during Implementation when they choose to pursue too many innovative ideas at once. By committing to the simultaneous pursuit of too many ideas, companies become overcommitted in terms of resources. Also, people are being asked to focus their attention in too many directions at once.

As an innovation facilitator, I hate to see any idea die. But without an effective method for selection of ideas that move forward and those that don't, companies tend to run into the Focus/Resources Bottleneck.

As I mentioned earlier, every innovation program should have a portfolio of ideas being pursued. Many like to describe this objective as if it were a funnel, with lots of ideas being poured into the top and only the best ones flowing out of that very narrow opening at the bottom. To achieve success, the team must periodically make some tough decisions about which ideas will move forward. It must also know when to stop, even if it's at the last minute.

If you're getting trapped in the Focus/Resources Bottleneck, this should become clear as you develop your implementation road map. At that point, you may need to step back and decide if some ideas need to be put on hold temporarily.

Sometimes, however, overcommitment may not become clear until you've already moved into Implementation. The tag team sessions that I'll talk more about shortly should help you uncover the Focus/Resources Bottleneck and allow you to take appropriate action.

Quantity of Ideas versus Quality

There is a serious negative by-product from the Focus/Resources Bottleneck. I call it Innovation Imbalance, and it arises when an organization, in its desire to be innovative, inadvertently chooses one of two extremes:

1. To pursue so many incremental ideas that it can never take the time to marshal its energies behind the thoughtful development of a potentially breakthrough one, or
2. to pursue only one potentially breakthrough or transformational idea at the expense of all other ideas.

This is a very delicate balancing act. In *Built to Last,* Collins and Porras celebrate the enormous risk taken by Boeing when it, quite literally, "bet the farm on the 747." While its risk paid off, it could just as easily have gone the other way. I would argue that this kind of innovative brinkmanship, while occasionally inspiring and successful, is not a healthy strategy for any company's long-term survival.

At the other end of the spectrum is the company that is attempting to move 50 innovations through its internal development systems. Whenever I see a company that tells me it is pursuing more innovations than it can handle, I know it is an incrementally driven innovator, at best.

Many companies, particularly in the consumer goods arena, deal with this challenge by assigning the incremental innovations to the existing businesses that manage them day-to-day while setting up a separate group responsible for managing the breakthrough opportunities. An interesting dynamic usually results from this approach and, too often, it's not a pretty one because it is the ultimate not-invented-here story. It's also one I've lived through more than once.

Quite simply, autonomous new venture groups tend to threaten the security of those who run the established lines of business. I'm not sure why this is true; I just know that it is. And while some of it has to do with competition for the same resources (further exacerbating the Focus/Resources Bottleneck), it's a lot more than that.

I've actually participated in meetings between new venture teams and people from the established business side where grown-ups act like children in front of their managements, pointing fingers and lobbing verbal bazooka blasts and other kinds of weaponry at each other. In one project, a team created a fairly innovative concept for a $40,000 investment, only to be forced into conducting over $250,000 in market research and quantitative testing because the division vice president, who would be inheriting the idea, fought its development every step of the way. I then watched him attempt to take partial credit for the success of that same idea once it was moved to his group for implementation.

I often describe the hand-off that needs to occur between the group that develops an idea and the broader team that is charged with implementation and launch as a relay race. Once, when I was talking about this in front of a client group, someone held up his hand and said, "You haven't described us." I asked what he meant. He said, "We don't hand the baton off to each other; we have to throw it over a wall." And someone else chimed in with, "Yes, and sometimes it comes back over the wall," and another person added, "And when it comes back over the wall, it's on fire!"

When I share this story with prospective and new clients, it is very telling to me whether they chuckle or whether this anecdote is greeted with silence. Healthy cultures—ones that will do well during the Implementation and Launch phase (because the level of animosity is low and the spirit of cooperation is high)—know how to conduct a smooth hand-off.

Unfortunately, the human dynamic between groups of people in any organization is sometimes driven by things other than what's right for the business. Strong leadership and a healthy corporate culture will minimize but not totally eliminate such problems. Constant vigilance is required to make sure that some individuals' personal goals are not interfering with the accomplishment of the organization's innovation goals.

The Bataan Death March Roadblock

As you'll probably gather from the fact that I named this roadblock after the deadly jungle march Allied prisoners were forced to make after the Japanese captured the Philippines, the Bataan Death March is one of the worst scenarios I've ever seen play out during an Implementation effort. In some organizations, the drive toward innovation can become so all-consuming that people and careers are mowed down in its path.

Generally, this kind of situation develops in organizations that have fallen so far behind their competitors that they develop a mentality that says, "This innovation is going to save us and so we must do whatever it takes to get it done." Such thinking is linked to the home-run myth that I warned against in Chapter 1.

Envision a place where people are forced to work long hours for interminable periods of time, punished severely for any mistakes that slow down the innovation juggernaut, and where resources are drained from every other part of the company to be poured into the innovation effort. Picture also a situation where the long-term overall health of the organization is the last thing on the minds of those who are wielding the power and who, too often, view the innovation effort as a potential career maker for themselves.

This is very dangerous stuff. Even if the innovation effort succeeds, the damage left in its wake can be staggering. Good people leave and those who are left behind are so worn out that the last thing they want to do is hear the word innovation ever again. And so begins another period in which the company ignores innovation until it absolutely can't afford to ignore it any longer.

The way to avoid this disaster, of course, is to build an organization that continuously innovates and, as a result, never has its back to the wall. But if you've failed to do that, you must avoid the Bataan Death March approach at all costs. It simply isn't worth the price.

The *Marty* Dilemma Roadblock

The final problem to be alert for as you develop your Implementation road map is the *Marty* Dilemma. This name refers to the oft-quoted dialogue from the 1955 movie *Marty,* for which Ernest Borgnine won the best actor Oscar. In many scenes of the movie, the interaction between Marty and his do-nothing friends goes something like this: "What do you want to do, Marty?" "I don't know, what do you want to do?" "I don't know, Marty, what do you want to do?"

In organizations that rely too much on consensus and not enough on Championed Teamwork, the exercise of developing an Implementation road map can become very much like that movie dialogue. If decision makers don't take a strong leadership role, critical choices—such as which innovative ideas to implement first and which resources will be deployed—get bungled.

Effective decision making at this stage can make or break an innovation effort. This is not the time for the leaders to leave the room and let the troops hammer things out by themselves. Decision makers need to be fully engaged in the Implementation phase.

Turnover—A Controllable Innovation Killer

Decision makers aren't the only ones who need to be onboard during Implementation and Launch. Nothing kills an innovation initiative faster than a turnover in key personnel.

Here is where I believe the requirements for leading effective innovation differ from those for running a visionary company. In *Built to Last,* Collins and Porras debunk a powerful myth that visionary companies are only built by charismatic founders. In truly visionary companies that survive for over 50 years through several leaders, a charismatic visionary founder is not the key to success. Some world-class companies, like 3M, have never had a famous, visible leader. Collins and Porras define the visionary companies as being those that have leaderships who craft long-term strategic visions for their companies that encompass the core ideology/envisioned future elements discussed in Chapter 7.

When it comes to innovation, my experience has been that there is still a need for a visionary (not necessarily charismatic, but it helps) champion. And, while I can't document this belief as scientifically as Collins and Porras did for visionary companies and leadership, I will always choose the champion over group consensus any time.

For this reason, turnover can kill even the best, most innovative idea. Often what is of critical importance as you prepare your road map and begin to move through Phase 5 is one true champion for the idea—one individual who believes in the idea as much as Art Fry at 3M believed in his Post-it Notes idea and as much as Lionel Pilkington believed in his floating glass theory. Because the world is short on champions of this caliber, you need to do everything possible to ensure that if you have such a person on your team, that person stays with the project through the implementation phase. Here are some problem areas to watch out for regarding this impediment:

- In too many organizations, rotating team assignments have become the norm, and people are moved off of projects without consideration for how much the project team will lose if that person has been a champion of the idea being pursued. I am constantly flummoxed by the seemingly irrational game of musical chairs that most senior management teams seem willing to play, especially on allegedly critical innovation projects.
- As you pull in the new people who will work with your innovation team during the Implementation effort, be aware of the signals you're sending as you choose the new team members. In some organizations, weak people are pulled away from the day-to-day operation, where the company makes its money, and thrown into innovation Implementation teams. Do this often enough and you'll find that strong

people are reluctant to work on Implementation efforts because innovation is perceived as a career graveyard. This becomes a self-fulfilling loop—weak Implementation teams lead to weak innovation results, which means that the next Implementation effort will draw weak people and produce weak results.

- If one or more of your champions moves on to another company in the middle of an Implementation effort, be sure to find out why. Too often, it isn't for the money; it's because they got frustrated with the inability to get their project through the pipeline. A series of such defections signals a major problem in your culture that needs to be addressed before innovation can succeed.

Never underestimate the impact that culture has on people pursuing innovation. Bad cultures can drive off champions, and a great, innovation-supporting culture can turn even Casper Milquetoast into a committed champion. One of the very best clients we've ever had was an absolutely miserable client when we first worked with him at a previous company. He was so bad that when he moved to his new company and called us to do some work there, I almost turned him down. But we did accept the work, and found ourselves dealing with a totally new person—he was a superior client and it was because he was now in a culture where innovation was allowed to thrive. He was able to assemble a great team. He was truly empowered to make decisions. And he knew what business he was in so that he didn't need to constantly seek approval for his every move.

Tag Team Sessions

Decision makers continue to have a role to play as the implementation effort moves forward after road mapping. They should participate with the original innovation team in periodic meetings where the team checks in with the people who are in charge of the key implementation functions. We call these tag team sessions. Regardless of the specific process you might use, these meetings are more of those shirt-sleeves sessions that become critical both for interdisciplinary teamwork and interactions with senior management decision makers.

Tag team sessions are where big and small issues are handled. They can range from the macro to the micro and always require cross-functional teamwork, collaboration, and a skilled facilitator.

Do not fall into the trap of thinking that communication via memos and e-mail can replace these sessions. Nothing can supplant the high value of the face-to-face, frank exchange of information and viewpoints that tag team sessions provide.

Here are some dangers to watch out for in your tag team meetings:

- Be sure these meetings don't degenerate into finger-pointing sessions. Use the reflective idea evaluation skills your team learned in the Discovery phase to problem solve in a way that allows people to resolve differing points of view.
- Don't let the sessions turn into opportunities to throw the yes/no switch. Again, the team's idea evaluation skills should enable you to find ways around obstacles without allowing the idea slayers to dominate and kill off ideas.
- Here again, team turnover can cause problems in these sessions because, as people leave, they may take parts of the institutional memory of the innovation effort with them. For instance, the reasons behind key decisions that were made during the Greenhouse phase may be forgotten. Also, as new members join a team, they may be less likely to be champions for the effort because they didn't participate in its earlier phases; not having participated in the birth of the idea, they don't feel any parental obligation toward it.

These problems can cause significant hiccups for the innovation effort. You need to be aware of them and address them in your check-in sessions.

Touchstone Sessions

We refer to periodic linkages back to the core ideology and envisioned future as touchstone sessions. They are the only way to measure how on track or off track your innovation initiative might be.

For example, a close friend once worked on the test marketing of a revolutionary new shampoo technology, which included a patented ingredient that actually used any available light to improve the shine of the hair. This R&D-driven concept did extremely well in consumer testing and in a regional test market.

Unfortunately, large-scale manufacturing issues made the original R&D formula unworkable for national launch. Nevertheless, the division's senior management instructed R&D to reformulate for national launch.

Now, here's where the touchstone part comes into play. Periodic revisiting of the original premise for this product should have demonstrated that the patented ingredient was the reason for the product's existence. Removal of it negated the product's entire reason for being and should have caused it to die. But, because the innovation juggernaut was on the march, the idea proceeded toward certain doom. The original marketing plan, which had worked so well in test, resulted in the expected high level of trial for the product. But the product no longer delivered on its original performance promise so it produced very low repeat purchases. This outcome would have been predicted if the innovation effort had remained closer to its original strategic touchstones. Instead the goal shifted to get something out the door.

Learning from Our Successes and Our Failures

In eighth grade, I had a Social Studies teacher, Mr. Resnick, at Webster School in Manchester, New Hampshire. One day he spent an entire class period helping 25 14-year-olds understand the meaning of the statement, "What we learn from history is that we don't learn from history." And, while it has profound meaning for the whole world and the way humankind continues to make similar mistakes (remember the Machiavelli quote), it also has an important message for innovators.

How many companies truly document and learn from their ongoing innovation initiatives? In my 30-year career, I have come across a handful. How unfortunate, because instituting a debriefing program that extracts the key learnings from each innovation effort and uses them to improve future efforts is invaluable. If conducted in the right way, which means in the spirit of joint learning and improvement rather than blame and punishment, this debriefing can clear the way to smoother rides on the innovation highway.

A roadblock that can occur at this point is the invocation of the Onward Mantra. Some organizations are loathe to look back and evaluate how they've done; they just want to move onward. Generally, this behavior arises out of a lack of understanding of how to look back without such a process turning into an exercise in blamestorming—where only the bad and none of the good is recognized.

Similarly, some companies fail to celebrate their innovation successes or to appropriately reward those who contributed to them. This can cause good people to look elsewhere for opportunities where their efforts will be recognized.

Finally, and most importantly, reward the well-constructed failure. Nothing sends a stronger signal to the organization that management understands the risks associated with innovation than the celebration of failure. And, if you become adept at extracting learnings from every experience, like Tom Edison (no failure, only learnings), you will have fostered the most innovative kind of culture possible.

So take time to celebrate both successes and failures, and take time to consider how things could be improved in future innovation efforts. Both activities will prove valuable the next time you travel down this highway.

[i n n o v a t i o n   f u e l]

- If you are pursuing a really breakthrough idea, be prepared for a long stay in Phase 5, Implementation and Launch. Also be ready for Dark Night of the Innovator episodes that will test your organization's true desire to be innovative.

- Be sure that your road map for implementation takes into account how departments and individuals will have to work together to achieve success. Do not ignore past antagonisms; deal with them head on so that they don't derail your innovation effort.

- Avoid the Focus/Resources Bottleneck roadblock. Have the courage to put some strong ideas on hold while you pursue others. Do not overload your innovation plate to the point where people and resources are exhausted. If you do, it will probably mean you are pursuing too much incrementalism and not enough breakthrough.

- Avoid organizational structures that pit innovation teams and existing business teams against each other.

- To the greatest degree possible, keep personal career objectives from interfering with the innovation effort. Be alert for individuals who want to sabotage the implementation effort as well as for people who try to drive an unwise initiative forward for reasons that do not match the organization's best interests. Above all, do not send your team off on the corporate equivalent of the Bataan Death March.

- Be decisive; don't fall into the *Marty* Dilemma roadblock where nothing gets done because no one steps up to the plate and champions the innovation cause.

- Don't let your innovation effort be hampered by personnel turnover. Identify idea champions and do everything possible to keep them on board and involved.

- Use tag team sessions (where decision makers get to check in with the innovation team) and touchstone sessions (where you make sure that you're still in touch with your original objectives) to keep things moving in the right direction and strategically aligned.

- Learn from your mistakes and your achievements; celebrate and reward successes and well-run failures.

CHAPTER SIXTEEN

Final Thoughts—Five Essentials of Innovation

Our travels together on the road to innovation are coming to an end. It is now time for you to embark on your own trek down the innovation highways and byways that will lead to exciting new products, new services, and other new ways of doing business. I hope what you've learned here will help you smooth out speed bumps and avoid roadblocks throughout your trip. I also hope you're able to make all types of bazookas a thing of the past in your organization.

As you put the ideas and techniques in this book to work, there are five essentials for innovation that you should keep in mind, no matter where your innovation journey takes you. These are attitudes and behaviors that will see you through the Dark Night of the Innovator and will make even the best-planned and best-executed innovation processes easier to pursue. I've touched on all of them in one way or another in earlier chapters but they are so important that they bear repeating. In my experience, no innovation effort can reach its full potential without these five innovation essentials:

1. Cross-functional communication
2. Persistence
3. Trusting your educated gut
4. A willingness to try many paths
5. Learning from failure

"What We Have Here Is a Failure to Communicate"

If death certificates were written for failed innovation efforts, the cause of death frequently would have to be listed as failure to communicate. Good communication is essential in every phase of innovation, and it is particularly critical at those points where various parts of your organization—including ones that have little history of interaction with innovation efforts or have a record of negative interactions—have to come together to move your process forward.

At every step along the innovation highway, double-check to make sure the channels of communication between the necessary parties are open and operating smoothly. Be certain everyone understands the strategic vision and keep coming back to that in your communication as the touchstone around which all else revolves.

As you move through the various innovation phases and need to expand your innovation team, thoroughly communicate to new team members what has happened in previous stages and what their roles will be as you move forward. This sharing of information and expectations will help galvanize the newly expanded team into a cohesive unit.

Be alert for any signs of internecine warfare between the various functional units that are involved in your innovation effort. Past differences can be put aside and strong new alliances can be formed if communication is handled properly. Sometimes, this may require the intervention of the decision makers to end turf battles. When things seem to bog down because of cross-functional differences, leaders must reemphasize their support for the strategic vision and for the innovation effort as an important step toward that vision.

Fill Your Suitcase with Resolve

As many of the stories I've shared with you have illustrated, steadfastness in the face of seemingly insurmountable challenges is the hallmark of a true innovator. As Albert Einstein said, "I think and think for months and years. Ninety-nine times, the conclusion is false. The hundredth time I am right." Organizations that aren't willing or able to get to that hundredth time are the ones that end up as casualties along the innovation highway. Make "Persistence pays!" your mantra, and it will guide you through to success.

Where Insights Come From

Trusting your educated gut is exceedingly hard for some people and for many organizations. If you are in a company that lives and dies by market research numbers, it will be difficult to convince people that following your collective intuition is often a better way to achieve innovation. Yet time and again, I have seen this proven. And, as I stated earlier, the astronomically high failure rate of allegedly fully tested new products and services alone makes a strong statement about the ineffectiveness of basing everything on the numbers and ignoring what your educated gut says.

The good news is that going with your educated gut is habit forming. Once you've had some success with trusting the insights that come from your collective wisdom, it becomes easier to do it again. Making this cultural change may take a while, but if you are persistent (there's that word again), it can be done and greater success will follow.

There Is No One Way

I've said these things before but I can't emphasize them enough: There is no one right way to innovate and no one can provide you with a road map that is guaranteed to lead you to success. You must be willing to explore many paths and to veer off on detours that your gut tells you might be interesting. Yes, detours can be time-consuming and you might waste some gas (and run into some dead ends), but remember that traveling only on interstate highways is generally mind numbing and rarely leads you through magnificent scenery. Besides, that's where everyone else is driving and the possibility of finding truly breakthrough ideas out in that heavy traffic is limited.

Just as I've advised you to fill your innovation portfolio with a variety of ideas ranging from incremental to breakthrough, I encourage you to be willing to venture off the beaten path in pursuit of those ideas. Visit other worlds (i.e., other industries) and snoop around in parts of your own organization where you don't normally roam. I guarantee that down one of these paths you will come across something unexpected and extremely valuable.

Value Mistakes

Through the centuries, untold numbers of innovations have resulted from what initially appeared to be mistakes. If your organization has any hope of being innovative, it must accept mistakes and, equally important, learn from them. Leaders who insist that their people operate mistake-free have no business on the innovation highway—and, sooner or later, they will have no business at all.

Accept the wisdom of legendary jazz saxophonist and composer Ornette Coleman who is credited with developing atonal, free-form jazz. As Coleman put it, "It was when I found out I could make mistakes that I knew I was on to something." Innovation is all about mistakes. You'll get it wrong more often than you'll get it right. But if you learn something from each wrong step and each dead end you eventually will get it right.

You've now fueled up your tank for a trip on the innovation highway. You know what roadblocks and speed bumps to expect and how to smooth them out. Above all, remember Frank Hine's favorite piece of advice from Miss Frizzle, of *The Magic School Bus*, that I mentioned in the Introduction: "Take chances . . . get messy . . . make mistakes."

[i n n o v a t i o n   f u e l]

- Communicate, communicate, communicate.

- Be steadfast.

- Tune into your educated intuition.

- Follow many paths.

- Mine your failures for lessons for your next journey along the innovation highway.

Postscript

You now should be well equipped for a successful trip down the innovation highway. At the same time, of course, you are more aware than ever how challenging such a journey can be. As I wave goodbye, I would encourage you to make a habit of reaching out to others who are traveling on the same road. Every time you get to explore how other companies are tackling the tough issues of innovation is a chance to learn some useful tidbits that might improve your own organization's chances of success.

Make learning about innovation your quest. Having done so myself, I can bear witness to the excitement and fulfillment that seeking new ideas and making them real can bring to your life.

If, on your journey, you run into hazards that you don't know how to overcome based on what you've learned here, or if you just want to discuss some aspect of innovation that still puzzles you, please e-mail me at mark@creativerealities.com. I am always happy to hear from my fellow travelers on the innovation highway. Good luck and bon voyage!

Ground Rules for Effective Interaction

Establishing effective ground rules at the start of any innovation-focused meeting goes a long way toward eliminating innovation speed bumps. At Creative Realities, we post these rules and review them verbally with teams at the start of every meeting we facilitate. Do not assume that if you just review the rules once, everyone will instantly change his or her behavior; these lessons require constant reinforcement to effect true change.

Here are several of the ground rules we use in our sessions. Please remember to only use a maximum of five or six in any one session.

- *No bazookas!* Don't allow people to shoot down ideas; have toy bazookas, nerf balls, or other non-pain-inflicting devices that people can throw at bazooka wielders.
- *Start with a headline (keep it short).* Encourage people to share their ideas in the journalism mode, beginning with a headline of their point, versus the joke mode, which usually ends with a punch line after a ramble. If someone starts to ramble when presenting an idea, ask him or her to get the idea down on paper in headline form; move on to the next person with an idea and then be sure to come back to the first person.

- *No side whispers (one meeting at a time).* It is critical that people pay attention to the ideas that are being tossed out. Do not let distracting side conversations develop. They tend to suck the energy out of a room and create an in-crowd mentality.
- *Use a pad to capture ideas (doodle!).* Because everyone cannot talk at once in a well-run meeting, participants should jot down every idea that pops into their heads, either in words or as doodles, whichever format comes naturally to them. Ideas are fleeting and they need to be put down on paper to avoid the risk of losing them.
- *Build on others' ideas (steal, but give credit).* Thomas Edison once pointed out that no one ever came up with a great idea all alone. Building on other people's ideas tends to make them stronger and promotes a collaborative spirit. But, when team members are building on each others' ideas, it is essential that credit be given to the ideas that trigger those builds; otherwise, the people with the original ideas feel ripped off and start to mentally drop out of the meeting.
- *Offer ideas versus questions (questions mask ideas).* Instead of allowing people to ask questions, ask them what idea is behind the question. People who are tentative about putting forth their ideas will often disguise them in the form of questions.
- *Speak for yourself.* Encourage people to say what's on their own minds and not bring into the room what they assume may be the thinking of others in the organization.
- *Decision makers offer ideas, too.* Decision makers have to model the behaviors they want other team members to exhibit. Sitting quietly while waiting for others to be creative, in the mistaken belief that it will encourage subordinates to participate, is a surefire formula for failure.
- *That was then . . . this is now.* Don't allow people to get bogged down in past successes, past failures, or "the way we've always done it."
- *Speak now . . .* This reflects upon the trust level of the group, by challenging everyone to share whatever is in their minds openly in this forum, as opposed to later, in one-on-one minimeetings with the decision makers, as sometimes happens in unhealthy cultures.
- *Stay loose till rigor counts.* This is one of my favorite George Prince quotes, so I've turned it into a ground rule. There are "seasons of thinking" when developing new ideas, some involving divergent thinking and some requiring convergent thinking. Evaluation is a convergent function. Stay open to all possibilities until you reach the point where final decisions need to be made.

- *Think and link.* Look at the ideas presented to see where there are opportunities to bring two or more concepts together.
- *Trust your educated gut.* If an idea is truly intriguing but on the surface seems undoable, don't let it go. Your educated gut is telling you something; follow it.
- *Model Championed Teamwork.* Designate a facilitator who can enable both teamwork *and* entrepreneurial decision making, thereby eliminating lowest-common-denominator consensus in favor of the more productive I-can-live-with-it kind of collaboration.

Sample Controlled Mental Flights of Fancy

These exercises have a place in every phase of your innovation journey. They are critically important in the Discovery and Invention phases, but they can also help you find solutions in every other phase, too. These techniques can also be used in other parts of your business outside of the innovation process. Any time you need to come up with some new ideas, find a new solution to a vexing problem, or reenergize a group of people who have been struggling too long without finding the answers they need, try one.

Hundreds of these exercises exist and, after some experience, you'll be able to make up your own. Here are instructions for five that have served us well over the years.

Exercise 1: Worst Idea/Trash It

Gets the group's energy going by rapidly moving them into the absurd!

1. Setup:
   • Materials needed are a small waste basket and lots of blank sheets of paper, but in a crunch can be done without any materials at all.

2. Implementation:
   - Ask everyone to come up with his or her worst possible idea or wish that addresses the task. Look for terrible, completely absurd, awful ideas, the worst you can think of. (Give an example. Push for really bad; e.g., cement soup, stone mattresses, exploding computers.)
   - Allow a little time for people to write down their ideas and have each participant write one worst idea on a clean sheet of paper. Make sure it is legible.
   - Ask one person to share his or her wish with the group. Agree that it is perfectly horrible, then produce a waste basket (you can cover it with a bag in advance). Ask that person to crumple up that awful idea and trash it. If time allows, ask for another and another. If not, just tell the group you're sure their ideas are equally horrible and have them get up and throw their ideas in the trash can.
   - Tell them: "It's time to take out the trash. In fact, let's recycle this garbage." Pass the can around and have each person take out one worst idea (they can exchange if they pick out their own).
   - Ask "How can you turn this into a good idea?" (Model one with group; two steps.)
     - Probe 1: "As bad as this idea is, is there something of value in it? How can we take that and make a good idea?"
     - Probe 2: "Can you find a reverse or opposite of this idea, which we could use to create a good idea?"
   - Allow pad time to generate wishes (to task). Ask participants to first offer up the wish, then read aloud the worst idea that led to the wish.
3. Benefits:
   - Gets absurd fast. Instant stretching.
   - Gives permission to suspend judgment.
   - Creates lots of laughter and energy.
   - Fun for the facilitator, too.

Exercise 2: Color Fields

Here is a fresh perspective/nonverbal/break from linear thinking.

1. Setup:
   - Divide the group into small teams (two to four people each)

- Provide a "pile" of art materials (crayons, paints, glue, scissors, colored paper, pipe cleaners, etc.)
- Give each team a flipchart
- Some background music is helpful

2. Implementation:
   - Instruct each team to construct a work of art using any of the materials provided, as well as anything else it can find around the room that can be used without being destructive.
   - The one extra rule is that they cannot talk to each other.
   - After they have created their work of art, they each take a writing pad, and walk through the "company museum of fine art." As they walk, they make notes of things they like, impressions they get, whatever comes to mind.
   - Then, at their seats, they make a list of absurd wishes, stimulated by the artwork, that address the task or topic of the moment and share them with the group.
   - Finally, move from absurd wishes to connections to the task, with more realistic wishes.

3. Benefits:
   - Forces thinking completely outside the box, and consideration of the task from alternative perspectives.
   - Nonverbal exercises tap other communication skills

Exercise 3: Story Builder

Here is another fresh perspective/break from linear thinking/building on others' thoughts.

1. Setup:
   - No materials required. This exercise can be done in their seats or standing.
   - Instruct everyone to participate with a pad and pencil, jotting notes or doodles of any quick thoughts as the story proceeds.

2. Implementation:
   - Begin a story, about anything. Proceed along a fairly predictable path, then make it take a sharp left turn.
   - Pass it on to the next person to add a bit.
   - Continue passing until everyone has participated.

- Ask everyone to think about what they heard, pick out a part or parts, and force an association to the task.
- Ask for ideas based on this association.

3. Benefits:
   - Very collaborative, campfire-like feel.
   - Vivid storytelling promotes imagery.

Exercise 4: Different Worlds

Here is another fresh perspective/break from linear thinking.

1. Setup:
   - May be done as a full group or in split-group teams.
   - May be done in real time, on note paper, or on flipcharts.
2. Implementation:
   - Instruct each person or group to consider the task/challenge as if it were a similar issue, but he or she existed in a different world.
   - Each person is to provide wishes or ideas based on the needs of that world.
   - "World" ideas can be anything that seems relevant and may include:

The sea/fish	Entertainment
The jungle/animals	A family gathering
Nature (spring, rain, fire, hibernation, sleep, leaves)	*National Enquirer* headlines
A related business	Politics (democrats, republicans, communists)
A very different business	Martha Stewart
The home/family	Saturday Night Live
Philosophy	Art
Religion (stories from the Bible)	Energy
Clothing design, high fashion	Law
IRS	Banking
Food, wine, cooking	Government
Publishing	Mystery novels, who-done-its
Retail	Outer space
Marriage	Agriculture
Machines, cars, motorcycles	Old West
History (or U.S. history, Russian history, 18th-century Europe)	Music, rock-n-roll
Bacteria/micro-organisms	Greek mythology

- Make connections from the thinking generated in another world to the issue at hand. (Link it back initially with absurd wishes.)
3. Benefits:
   - Breaks the group out of a rut.
   - Introduces a perspective and ideas that would not normally be developed in a literal way.

Exercise 5: Whip It/Morph It

This is designed to help look at ideas from new perspectives.

1. Setup:
   - Below are instruction words; write these instruction words on individual index cards or copy the pages and cut the instructions into strips.
   - This exercise can help develop an idea or be used against the original task focus. A tangible challenge works best.
   - The exercise works best with pairs.
2. Implementation:
   - Give each participant an index card or strip of paper with an instruction.
   - Have each participant take the task or developed idea and adapt it according to instruction words. Have him or her generate as many options as possible.
   - Advise participants not to be too literal in interpreting their instructions—play with it!
   - Next, have participants pair up and share what they generated. Together, have them develop two to four wishes, "we coulds," or "what ifs" (depending on session objectives).
   - Share ideas generated; look for builds.
3. Benefits:
   - Looks at beginning idea or task from different perspectives.
   - Excellent for guiding the group into areas that would otherwise be overlooked.
4. Instruction Words:

Whip it	Fold it
Heat it	Cushion it
Roll it	Smash it

Lower it	Make it larger
Expand it	Make it fatter
Clamp it	Make it slimmer
Heat it	Make it wider
Slice it	Make it colder
Balance it	Make it hotter
Tighten it	Make it harder
Raise it	Make it softer
Bend it	Make it more flexible
Inject it	Make it more rigid
Freeze it	Make it smellier
Compress it	Make it more streamlined
Steam it	Make it brighter
Loosen it	Make it dimmer
Smear it	Make it smokier
Sink it	Make it more humid
Match it	Make it more arid
Smoke it	Make it fuzzier
Fry it	Make it thicker
Make it longer	Make it more fluid
Make it shorter	Make it spicier
Make it smaller	

Index